THE BIRDWATCHER'S
GUIDE TO HAWAI'I

DISCARD

DATE DUE

THE BIRDWATCHER'S GUIDE TO HAWAI'I

RICK SOEHREN

A Kolowalu Book

University of Hawai'i Press, Honolulu

02 03 04 05 06 07   8 7 6 5 4 3

Library of Congress Cataloging-in-Publication Data
Soehren, Rick, 1958-
The birdwatcher's guide to Hawai'i / Rick Soehren.
p. cm.
"A Kolowalu book."
Includes bibliographical references (p.) and index.
ISBN 0-8248-1683-8 (paper : alk. paper)
1. Bird watching—Hawaii—Guidebooks. 2. Birding sites—Hawaii—
Guidebooks. 3. Hawaii—Guidebooks. I. Title.
QL684.H3S64   1996   96-14596
598'.07234969—dc20   CIP

All photographs were taken by the author.

Cover painting of 'Apapane on 'ōhi'a by Patrick Ching.
Photo references: Jack Jeffrey (foreground) and
David Muench (background).

University of Hawai'i Press books are printed on acid-free
paper and meet the guidelines for permanence and
durability of the Council on Library Resources

Designed by Kenneth Miyamoto

To My Dad

Contents

Color photographs follow page 84.

Maps and Tables

Acknowledgments

THIS BOOK would not have been possible without the help of dozens of individuals who were kind enough to share their wisdom and experience with me. At the State Department of Land and Natural Resources, special thanks go to Ron Walker on Oʻahu, Tom Telfer on Kauaʻi, Jon Giffin and Miles Nakahara on the Big Island, and Meyer Ueoka on Maui. At the U.S. Fish and Wildlife Service, I am very grateful to Jim Jacobi from the Mauna Loa Field Station, Richard Voss, refuge manager at Kīlauea Point National Wildlife Refuge, Kathleen Fruth Viernes at Kīlauea, Jack Jeffrey at the Hakalau Forest National Wildlife Refuge, and especially to Dick Wass, refuge manager at Hakalau. At Hawaiʻi Volcanoes National Park, thanks go to Larry Katahira. Heartfelt thanks to Paul Higashino at The Nature Conservancy of Hawaiʻi for providing a great deal of information and advice on birding spots and for special help at the Waikamoi Preserve. Ed Misaki, preserve manager at Kamakou, also offered excellent advice for which I am very grateful. Thanks also to Dave Hill at the Hawaiʻi Nature Center. Reginald David of the Hawaiʻi Audubon Society provided encouragement and very helpful comments on the manuscript. Much of the credit for this book goes to these dedicated and enthusiastic professionals. Any errors or inaccuracies that remain are my responsibility.

There are others too numerous to mention who told me of favorite birding spots, offered advice, did favors, showed great hospitality, and helped in so many other ways. All of them demonstrated the spirit of aloha for which Hawaiʻi is justly famous, and all have my thanks and gratitude.

Finally, special thanks to Don Heath for teaching me the value of good access for the disabled, to Christopher Carr and Marsha Prillwitz, who encouraged and inspired me, and most of all to my wife Debbie, who supported me and who sacrificed a great deal so that I could tramp around in Hawaiian forests, chasing a dream.

1
Hawaiian Birding Basics

BEFORE ANY HUMANS ever found their way to Hawai'i, the Island birds and the forests existed in harmony, almost as a single living thing. It was no accident that the slender bill of a honeycreeper matched perfectly with the tubular flower of a lobelia; the two evolved together for millions of years. The bushy crest of another honeycreeper was coated with pollen as it flew from blossom to blossom in 'ōhi'a trees, the bird apparently trading nectar for service as a pollinator.

When Polynesians discovered these Islands and settled, they brought great changes. But still, the lives of people and birds were intertwined. The early Hawaiians searched the rainforest for birds like the 'I'iwi, whose scarlet feathers they fashioned into magnificent capes for their royalty. Elsewhere in the forest, canoe makers seeking giant koa trees were shadowed by a bold and curious little bird that followed them through the dense tangle of greenery. They called this bird 'Elepaio, and considered it their guardian spirit.

The Island forests are still home to birds like these, birds that evolved in the profound isolation of Hawai'i and exist nowhere else on earth. The scarlet 'I'iwi, prized by the ancient featherworkers, still flashes its bright plumage in some of the mountain forests, and even today the curious little 'Elepaio approaches quiet visitors who enter its forest home. Other birds, with lilting names like Palila and 'Apapane and Koloa, also dwell on these Islands. They are living examples of the evolutionary magic that filled Hawai'i with life.

The places where these birds dwell are the quiet places of Hawai'i where few people venture. Birding in these uncrowded spots is a little like stepping back into the forests of long ago, to a time when a natural harmony embraced the land and newcomers had just begun to fashion these Islands into their own vision of paradise. Even at the edge of busy Honolulu, it is easy to escape from the crowds and hike forest trails that

An undisturbed Hawaiian forest

are strongholds for Hawai'i's native birds and plants. In these places you can still experience some of the balance and order that millions of years of island evolution have achieved.

In other places, Hawai'i has taken on a new face. People from all over the world have come here with plants and animals they knew at home. Prominent among these imports are the birds: pet birds, game birds, water birds, land birds. There are so many introduced species that the Islands sometimes seem like giant aviaries stocked with an unbelievable array of colorful and fascinating creatures. These birds add diversity to the gardens, parks, and fields where they live, while adding spice to the birder's checklist. Who isn't delighted to spot a White-rumped Shama or an Erckel's Francolin?

The relatively small number of bird species found in Hawai'i compared to many continental areas can help visiting birders avoid frustration. Even on a short visit to the Islands an avid birder can learn all the species and see many of them. For residents, the comparatively short list of Hawaiian birds can be a little disappointing when they read of mainland Christmas bird counts with totals well into three digits. But keep in mind that few other areas have the diversity of marine, forest, wetland, and upland birds found in Hawai'i, and nobody has an 'Amakihi, an 'Apapane, or an 'I'iwi!

Sadly, Hawai'i must also admit to a level of avian rarity that no mainland area can match. Human impacts have taken a heavy toll on Island birds: 40 percent of the endangered birds in the United States are from Hawai'i.

A fringe benefit of birding in the Fiftieth State is that you're surrounded by other diversions while you're looking for birds. Many birding spots also offer hiking trails, sea turtles, rare plants, erupting volcanoes, dolphins, whales, historical sites, or coral reefs perfect for snorkeling. From the sublime experience of seeking out birds in the rain forest to the novel sight of canaries and parrots flying free in city parks, Hawai'i offers birding at its best.

The Birding Spots in This Book

The State of Hawai'i is composed of 132 islands stretching across 1,500 miles of the central Pacific Ocean. There are eight major islands that make up over 99 percent of the land area. Among these eight, two are off-limits to birders: uninhabited Kaho'olawe, formerly used for Navy operations, and Ni'ihau, privately owned and generally not open to visitors. The remaining six islands are the focus of this book. In order of size, they are Hawai'i (referred to in this book by its informal name, the Big Island), Maui, O'ahu, Kaua'i, Moloka'i, and Lāna'i.

The rest of the state is composed of a string of tiny islands that extends in a line to the northwest. These are known collectively as the Northwestern Hawaiian Islands, the Northwest Chain, or the Leeward Islands. The best known, Midway Atoll, and most of the rest make up the uninhabited Hawaiian Islands National Wildlife Refuge. This refuge provides critical habitats for many species, including the Hawaiian green sea turtle, the Hawaiian monk seal, and a wealth of seabirds. Because of the fragile environment and the potential for disturbance, access to these islands and reefs is limited.

Every one of the birding spots described in this book offers something special: uniquely Hawaiian birds that don't occur anywhere else, called endemic species; seabirds that rarely come ashore; or maybe just a profusion of introduced birds in an urban setting. All the areas in this book are open to the public. That's an important consideration in Hawai'i, where much of the land is privately owned and closed to visitors. In a few cases you'll need to join an organized group or get special permission to visit a birding spot, and information is provided on how to arrange this.

The birds of greatest interest and those most likely to be seen at each site are mentioned in the description of the site. Special attention is given to seasonal occurrence to help visitors plan their birding trips. The Occurrence Tables in the appendix provide information on the relative abundance of each bird species at each site so you can tell quickly what birds you are likely to see on a particular island.

As a practical matter, the facilities at each location are noted, including visitor centers, rest rooms, telephones, food, drinking water, entry fees and hours, and opportunities for picnicking and hiking. The description of each site also notes wheelchair access and any barriers to birders with restricted mobility. The heading for each site description summarizes facilities using the recreational symbols shown in Table 1.

Other Publications You'll Need

Good maps and a field guide to the birds of Hawai'i are essential to help you find and fully enjoy the birding spots described in this book. For the birder, the best maps of each island are *Reference Maps of the Islands of Hawai'i,* published by University of Hawai'i Press. Separate maps cover O'ahu, Kaua'i, the Big Island, Maui, and Moloka'i-Lāna'i. They are packed with detail and information, are printed in full color, and cost only a few dollars apiece. These maps are widely available throughout the state at bookstores, souvenir shops, and even grocery stores. You can also order them directly from the publisher by phone or mail. See "Additional Reading" at the back of the book.

You'll want to have a field guide to Hawaiian birds as a companion to this book. The Hawai'i Audubon Society publishes a very good little full-color guide entitled *Hawaii's Birds* that is widely available in Hawai'i for

Table 1
Recreational Symbols

? Visitor Center	⑪ Food	$ Entry Fee
👫 Rest Rooms	🥛 Drinking Water	🏃 Hiking
☎ Telephone	♿ Disabled Access	⛺ Picnicking

about $10.00. Look for it almost anyplace postcards and souvenir books are sold. Visitors can even pick it up at the Honolulu Airport between flights. The book can also be ordered directly from the Hawai'i Audubon Society. See the appendix "Organized Hiking" and "Additional Reading" in the back of the book.

Recordings of bird sounds can be tremendously helpful to birders, particularly when it comes to identifying forest birds like those in Hawai'i that often can be heard but not seen. The Hawai'i Audubon Society offers a two-cassette set entitled *Voices of Hawaii's Birds,* a companion to *Hawaii's Birds.* See "Additional Reading" in the back of the book.

If you're a serious birder or a real bibliophile, check out *A Field Guide to the Birds of Hawaii and the Tropical Pacific* by Pratt, Bruner, and Berrett. This beautiful volume contains detailed information on each species and is illustrated with excellent color plates. Since it also covers the South Pacific, you'll get a tantalizing peek at the birds of other Pacific islands. The book can be hard to find in bookstores, even in Hawai'i. In larger mainland cities, try shops that specialize in nature items or that cater to birders. You can also order directly from the publisher, Princeton University Press. See "Additional Reading."

A volume that can be found on the bookshelves of legions of birders is Peterson's *Field Guide to Western Birds,* second edition. This old standby covers Hawai'i, but the information is over 30 years old and quite out of date. In those three decades some native birds have dwindled, while other introduced species have spread. You'll want more recent information.

The Language of Hawaiian Birding

Some of the words in this book may be unfamiliar to you, either because they are scientific terms such as *endemic* or because they are Hawaiian words. Visitors should learn a few Hawaiian words, such as the terms commonly used by local people when giving directions: *mauka,* meaning "toward the mountains" and *makai,* meaning "toward the sea." A glossary of important scientific terms and Hawaiian words is included in the back of the book.

The names of Hawaiian places and especially Hawaiian birds can be very difficult to remember and very intimidating to pronounce when they are unfamiliar. A basic understanding of Hawaiian pronunciation will be helpful in remembering the names of birds like 'Akiapōlā'au, 'Anianiau, 'Amakihi, and 'Ākohekohe. In order to pronounce Hawaiian words correctly, it helps to know a little about the language and alphabet. The Hawaiian language was not a written one until missionaries and other westerners arrived. They created a written Hawaiian alphabet consisting—then—of just twelve letters. Today, a thirteenth character is also used. The five vowels are pronounced as follows:

> *a* as in f**a**ther
> *e* as in b**ai**t
> *i* as in b**ee**t
> *o* as in b**oa**t
> *u* as in b**oo**t

The pronunciation of some vowels changes slightly, depending on whether they are accented or unaccented, and what their neighboring sounds are. In an unaccented syllable, *a* is pronounced closer to the vowel in English *cup,* and *e* closer to that in *bet.* Each of these vowels can be pronounced long, and these are marked with a *kahakō* (macron)—*ā, ē, ī, ō, ū.* Here, "long" means a change in duration, not quality, with the sound drawn out. The difference between *a* and *ā* is just as important as that between *a* and *e,* or *m* and *n.*

Certain combinations of two vowels are pronounced as diphthongs —that is, with the first vowel accented and the two sounds making up just one syllable:

> *ai   ae   au   ao   ei   eu   oi   ou   iu*

The eight consonants are pronounced as follows:

> *p* as in s**p**in
> *k* as in s**k**in
> *h* as in **h**ouse
> *m* as in **m**at
> *n* as in **n**o
> *l* as in **l**ie
> *w* as in **w**ear / **v**ery
> ' as in oh-oh (the sound between the vowels)

The last consonant, indicated by a single opening quotation mark, was not written at first, but it is an essential part of the alphabet. Called an 'okina, it represents a glottal stop, which indicates a momentary break in the word with the vocal cords stopping, and then releasing the air.

The letter *w* is sometimes pronounced as a [w] and sometimes as a [v]. There are no hard-and-fast rules for its pronunciation, except that there is a tendency to use [w] rather than [v] after *o* or *u* and at the beginning of a word. In the bird names 'I'iwi and 'Iwa, the *w* is pronounced as a [v].

Accent is predictable only in words up to three (short) syllables. In these words, the second-to-last syllable is accented. In addition, a syllable with a long vowel (marked with a macron) is accented no matter where it appears. In terms of accent, all longer words are made up of combinations of these shorter ones. In Table 2, periods separate longer words into shorter forms so that you can see where the accent falls.

To help readers pronounce longer words, the more recent editions of the *Hawaiian Dictionary* show the accent units for each entry.

For a more detailed guide to pronunciation, see *All About Hawaiian,* listed under "Additional Reading" in the back of this book.

Throughout this book, common or vernacular names of birds are used. For the most part, these are taken from the American Ornithologists' Union *Checklist of North American Birds (amended 1993)* and the *Checklist of the Birds of Hawai'i—1992.* Scientific names can be found in the *Checklist* or in field guides. The common names of endemic species are generally the names given to these birds by the early Hawaiians, except where the same Hawaiian name was apparently used for two or more birds now recognized as separate species, as in the case of several 'Ō'ō species. (Actually, it is probably our knowledge of the Hawaiian language that is deficient rather than the early Hawaiians' knowledge of Island birdlife. I have no doubt that Hawaiians knew the difference between the O'ahu 'Ō'ō and the Bishop's 'Ō'ō.)

In two cases I use Hawaiian names for species that are not endemic, because the Hawaiian names have gained widespread use locally: Kōlea for Pacific Golden-Plover (or Lesser Golden-Plover, as it was previously known), and Pueo for Short-eared Owl or Hawaiian Owl. A few birds found in Hawai'i are given different common names in some of the older published materials, and sometimes birds are known in Hawai'i by common names not used elsewhere. In addition, the A.O.U. has recently adopted new common names for several birds (the Creepers), recognized a new species (the 'Akeke'e, formerly Kaua'i 'Ākepa), and split the Common 'Amakihi into three species. Table 3 provides a list of birds that you may see referred to by other names.

The What, When, and Where of Island Birding

There are birds almost everywhere in Hawai'i. You can spot dozens of species while you walk city streets, relax at the beach, or visit popular

Table 2
Pronunciation of Some Hawaiian Bird Names

In the column on the right, short accented vowels are marked. Remember that all long vowels (those marked with a macron) are also accented.

ʻAkekeʻe	ʻÁke.kéʻe
ʻĀkepa	ʻĀ.képa
ʻAkiapōlāʻau	ʻAkía.pō.lā.ʻáu
ʻAkikiki	ʻÁki.kíki
ʻĀkohekohe	ʻĀ.kóhe.kóhe
ʻAlalā	ʻÁla.lā
ʻAlauahio	ʻAláua.hío
ʻAmakihi	ʻÁma.kíhi
ʻAnianiau	ʻAnía.niáu
ʻApapane	ʻÁpa.páne
ʻElepaio	ʻÉle.páio
ʻIʻiwi	ʻIʻíwi
ʻIo	ʻÍo
Kōlea	Kō.léa
Koloa	Kolóa
Nēnē	Nē.nē
ʻŌmaʻo	ʻŌ.máʻo
ʻŌʻū	ʻŌ.ʻū
Pueo	Puéo

Table 3
Alternate Bird Names

Name Used in This Book	Other Names Used
Dark-rumped Petrel	Hawaiian Petrel, 'Ua'u
Red-tailed Tropicbird	Koa 'e 'ula
White-tailed Tropicbird	Koa 'e kea
Masked Booby	Blue-faced Booby
Great Frigatebird	'Iwa
Brown Noddy	Common Noddy, Noddy Tern
Black Noddy	Hawaiian Noddy, White-capped Noddy, Hawaiian Tern
White Tern	Fairy Tern
Black-crowned Night-Heron	'Auku'u
Koloa	Hawaiian Duck, Koloa Maoli
Hawaiian Coot	American Coot, 'Alae ke'oke'o
Common Moorhen	Hawaiian Gallinule, 'Alae 'ula
Black-necked Stilt	Hawaiian Stilt, Ae'o
Kōlea	Pacific Golden-Plover, Lesser Golden-Plover
Ruddy Turnstone	'Akekeke
Sanderling	Hunakai
Wandering Tattler	'Ūlili
Nēnē	Hawaiian Goose
'Io	Hawaiian Hawk
Pueo	Hawaiian Owl, Short-eared Owl
'Alalā	Hawaiian Crow
'Akeke'e	Kaua'i 'Ākepa
'Akikiki	Kaua'i Creeper
'Alauahio	Maui Creeper
'Ākohekohe	Crested Honeycreeper
Hwamei	Melodious Laughing-thrush
Japanese Bush-Warbler	Uguisu
Red-billed Leiothrix	Pekin Nightingale, Japanese Hill-robin
Red Junglefowl	Moa
Japanese White-eye	Mejiro
Spotted Dove	Lace-necked Dove
Rock Dove	Pigeon
Zebra Dove	Barred Dove
Red-crested Cardinal	Brazilian Cardinal
House Finch	Linnet, Papaya Bird
Nutmeg Mannikin	Ricebird, Spotted Munia

attractions around the state. But some of Hawai'i's birds, especially the native ones, are restricted to very limited ranges. Many species, especially some of the little green forest birds, can be quite a challenge to distinguish because they are rather similar to one another. Success in finding and identifying these species depends on knowing what to look for, and looking for it in the right place at the right time.

Sometimes the good birding spots in Hawai'i are not well marked. You just have to know where to turn or where to stop. Mileage markers along the state's highways can help you find your way to birding spots. Between these signs, which are posted every mile along most routes, you can use your car's odometer to accurately measure tenths of a mile and find the spots in this book that are described by mile point.

In the forest, knowledge of different birds' habits will help you to find and identify them. For instance, Hawaiian forest birds tend to restrict themselves to particular layers of vegetation. Two fairly common endemic species, the crimson 'Apapane and the scarlet 'I'iwi, are most often seen at the very tops of dominant 'ōhi'a trees, feeding on the nectar of the blossoms. This top layer of the forest is called the canopy. The best viewing of these canopy-dwelling birds tends to be from spots

Forest canopy view from a canyon overlook

where you can look down on the treetops, such as on a hillside. Some other endemic birds like the 'Elepaio prefer the understory—smaller trees and shrubs that grow beneath the canopy. These birds can be seen best when you are inside the forest where the birds will sometimes approach you right at eye level.

Another Hawaiian tree used by many endemic birds is the koa. It occurs in many of the same areas as the 'ōhi'a, and like the 'ōhi'a it is a large tree that forms the forest canopy. Several forest birds, such as the bark-picking 'Akiapōlā'au, glean koa trees for insects. Koa flowers also produce a small amount of nectar. Koa seedlings bear a few feathery leaves, but as the trees grow they bear leaflike phyllodes, which are flattened stems, rather than leaves.

There are other clues you can use to identify species. Pay attention to coloration. Observe colors and patterns on the head and body, note any eye rings, and look for the color of the bill and legs. Bill shape can be quite important in distinguishing birds, too. Note how broad the base of the bill is in comparison to its length. To what extent is the bill curved? A bird's activity may also offer clues to its identity. Some species tend to creep along tree trunks, while others feed on nectar. Vocalizations can help you distinguish species once you've gained some experience listening to them. Finally, watch for associations with other bird species. Some Hawaiian birds will travel in mixed flocks.

When to look for birds can be as important as *where* to look. Some marine birds come to islands only to rear their young, so it is important to visit their colonies during nesting seasons, which vary by species. Many wetland species are migrants that spend only the winter months in Hawai'i. For some species, the "winter" stretches from August through April, while others may be present only during a few midwinter months. Forest birds can generally be seen year-round, although they may be more active and visible in the late winter and spring months during breeding season.

Early morning is usually the best time to see birds, especially forest birds, because they are most active and most vocal at that time. There are a few exceptions: at some sites your views of birds will be to the east and viewing is best later in the day when the sun is behind you.

Where to Stay

If you are visiting Hawai'i, you'll probably spend a fair amount of time around your lodging, so you might want to select it with birdlife in mind.

Some of the best places to stay are resorts with low-rise buildings of two or three stories, surrounded by lots of landscaped grounds. The landscaping tends to attract birds to the area, and you may be able to draw them in to your lanai or balcony with seeds, bread, or crackers. Depending on the island and location, you might attract Common Myna, Northern Cardinal, Red-crested Cardinal, Zebra Dove, House Sparrow, and maybe even a surprise or two. If you are on the ground floor, you may also attract mongooses, so be sure to keep the screen closed!

Low-rise resorts with lush landscaping tend to be among the more expensive places to stay, but there are economical alternatives. Several agencies in Hawai'i can arrange bed-and-breakfast lodging at small inns or the homes of local people. These accommodations are usually in residential areas where many urban birds are found.

Another lodging consideration is proximity to the best birding spots on an island. You may want to select lodging that is close to the sites that interest you so you can get into the field early.

Getting Around

The best way for the visitor to get to birding spots in Hawai'i is to rent a car and drive to them. Some of the places in this book are within walking distance from hotels, but many birding spots are not in tourist areas. Among the Hawaiian Islands, only O'ahu has a comprehensive bus system and it really isn't convenient for birders. Most sightseeing tours are not set up for birders either. They don't go to the good birding spots, or they get there too late in the day, or they don't stay long enough to satisfy a birder.

Nearly all the sites described in this book can be reached in a passenger car, but there are a few places in Hawai'i where you will need a four-wheel-drive (4WD) vehicle to get to remote birding areas. Many car rental companies prohibit the use of their vehicles, including 4WD vehicles, off the pavement. In some cases rental vehicles that look like 4WD really aren't, or else they are adjusted so the extra drive wheels won't engage. Be sure to check on these points before heading for remote and rugged areas.

At many rural parks and trailheads, you will see signs warning you not to leave valuables in your car. Heed these warnings, for auto breakins are common in Hawai'i. Although the signs warn you to lock your car, some people avoid broken windows by leaving their cars unlocked *with nothing of value inside.* Rental cars are the ones that get hit the most. The damage coverage offered by car rental companies is expensive and may duplicate your own insurance, but if you will be keeping the car only a day or two, or if you know you are going to an area where break-ins are common, the convenience is probably worth it. When you have a car more than a few days, the cost of damage coverage can really add up. If you decline this coverage, try to make your vehicle look less like a rental car. It may help protect you.

Being Prepared

The Boy Scout motto to "be prepared" is a very good idea. When you're in the field you'll need to protect yourself from the tropical sun, wind, and rain. When you're hiking you'll want the standard hiking gear, and there are a few birding items that you'll never want to be without. Here's a packing list:

The Birdwatcher's Guide to Hawai'i. Of course!

A field guide to Hawaiian birds. *Hawaii's Birds,* by the Hawai'i Audubon Society, is the best choice for the field because it is small and light.

Notebook. You'll want to keep notes on the birds you see. If you spot a rare bird, it is especially important to take detailed notes of the time and place you saw the bird, its appearance, sex, vocalizations, behavior, and the weather and lighting conditions. Be sure to use water-proof ink or pencil in this rainy climate.

Binoculars. Many birds in Hawai'i can be observed without them, but a pair of binoculars will be indispensable for viewing distant birds as well as whales, scenic points, and so forth. Compact binoculars are good for the traveler.

Hiking boots. If you do any hiking to see birds, you will want ap-propriate footwear. Hawaiian hiking trails can be very slippery when wet, so lugged soles are helpful. Slippery surfaces and uneven terrain can result in twisted ankles, so high-top boots and the ankle support they provide are a good idea. Lightweight, breathable, waterproof boots are ideal for Hawai'i's mix of heat, water, and rough terrain.

Canteen or water bottle. You can lose a lot of body fluid in the Hawaiian sun, and it is not safe to drink from streams. That waterfall may look pristine, but you can't see the dairy cows, feral pigs and goats, rats, and mongooses upstream that have deposited harmful organisms in the water. One of the biggest concerns is leptospirosis, a disease that can cause mild to severe flulike symptoms or even death. It is caused by a spiral bacterium that can enter the human body through the nose, mouth, eyes, or broken skin. Don't drink from Hawaiian streams.

Insect repellent. Mosquitoes can be abundant, especially in rainy lowland areas. At a few birding spots, specially noted in the text, you can't stand still very long without repellent. (Fortunately, many good birding spots in Hawai'i are at high elevations, above the range of resident mosquitoes.)

Sunscreen. The tropical sun can burn, and some birding spots offer little or no shade.

Sunglasses. Bright sun and sparkling ocean can make for a lot of glare. Polarized lenses are best for Hawai'i because they cut reflected light from water. (This is a good feature for peering down to see fish and other aquatic organisms, too!)

Hat or sun visor. These help keep stray light from your eyes when viewing birds, and also help prevent your forehead from getting sunburned.

Raingear. Even sunny days can see passing showers. Breathable waterproof fabrics are expensive but they can be well worth it in Hawai'i. If you plan to spend a lot of time in the wettest areas such as the Alaka'i Swamp, you may want rain pants. In a pinch, you can make a rain poncho by cutting arm and neck holes in a plastic garbage bag. Be sure to pack plastic bags for camera and binoculars, too.

Warm clothing. Most of Hawai'i's remaining endemic birds are found in high-altitude forests where it can be quite cool, and wind and rain can make it feel even cooler. Long pants and a jacket may come in handy.

Camera and film. You may be able to get some great photos of birds, especially with a telephoto lens of 200 mm or more. It can be dark in the deep forest, so you'll want fast film—ASA 200 or better. A lens-cleaning kit is a must for removing dust and salt spray.

Flashlight. You might stay out birding later than expected. Darkness falls more quickly in the tropics because the sun drops straight below the horizon, not at an angle as it usually does at higher latitudes. A flash-light will also come in handy for hiking through tunnels and lava tubes.

Snack or picnic lunch. Some birding trails are pretty strenuous. A light snack like peanuts or trail mix is good for quick energy.

First aid kit. This is always a good idea when hiking in isolated areas.

Daypack or fanny pack. This will carry all the other things on the list.

Brush. A small, stiff brush is useful for cleaning seeds off your boots and clothes before you leave a site, so you don't transport alien weeds to other areas.

Thermos. After an early morning walk through a cool misty forest, a cup of steaming Kona coffee is pure joy.

Caution. Even when you are well prepared, the accessibility of sites described in this book should be taken as a general guide only. Weather conditions can alter paths, slides can cover portions of trails, and so on. Always use common sense when out in nature, particularly in remote areas.

Taking Care of the Land

Much of the damage done to Hawai'i's environment has been caused by the introduction of alien species, because these introduced plants and animals often out-compete native species that evolved in an environment of little or no competition. Hikers can unwittingly transport the seeds of introduced plant species if they are not careful. One of the worst plant pests in Hawai'i is *Clidemia hirta*, or Koster's curse. This shrub is choking out native vegetation all over O'ahu, and it has been impossible to

Clidemia or Koster's curse

eradicate once it is established. The tiny seeds are spread by birds, but they can also stick to boots and pants. The yellow and strawberry guavas are also plant pests. The fruit may be tasty, but the plants grow so fast and so close together that they crowd out native plants and even obliterate trails. Another group of plants with tasty fruits and very bad growing habits are the brambleberries: blackberries and raspberries. These plants can grow into impenetrable thickets that crowd out all other life.

There are a few simple things you can do to avoid spreading these and other plant pests. Wash or brush your pants and brush your shoes carefully between hikes, especially when traveling between islands. Pay careful attention to shoelaces where seeds can stick. Some plant species first arrived in Hawai'i as seeds in the digestive tracts of birds. Remember that humans can transport seeds this way, too. Don't unwittingly plant berry or guava seeds along a trail. If you are visiting Hawai'i, make sure your gear is free of hitchhiking seeds before you come. Your efforts will help ensure that Hawaiian plants won't face additional unnecessary competition from introduced species.

2
A Hawaiian Natural History

OF ALL THE BIRDS to be seen in Hawai'i, the endemic forest birds generate the most wonder and enthusiasm. These are the creatures we can't see anywhere else—they are reflections of Hawai'i's powerful sense of place. When you're hiking in the forest, the koa and 'ōhi'a trees will tell you by their silent presence that you couldn't be in any other forest on earth, but the voices of a choir of endemic birds will insist that this place is unique, special, and irreplaceable.

As you marvel at these forests and these birds, you are bound to wonder how all of this unique life came to exist in Hawai'i. There are—or were—not just a handful of unique birds, but dozens of species. And their homes are in forests full of plants that are different from any others, anywhere. Tragically, these plants and animals are vexingly incapable of holding their own against the plants and animals we have introduced and the disturbance we have caused in the forests. How does such a unique and fragile system come into being? The answer lies in two of the most astonishing natural processes to be found on the planet: first, the geologic formation of this archipelago itself, and second, the remarkable evolution of the few plants and animals that initially made their way to these Islands. These processes explain the abundance of life that evolved in Hawai'i and show us why Hawaiian species and ecosystems are so vulnerable.

Islands from the Sea
Hawaiian geology and biology are profoundly interwoven. To understand how life arose on these Islands, we must first understand how the land itself was formed. The islands of Hawai'i have not always existed as they do today, but were created as a part of the process that shapes continents and the seas that separate them. These land masses and ocean

17

floors form the comparatively thin skin of our planet called the crust. This crust is fractured into numerous plates of varying size that move as the sea floor spreads along rift zones. These crustal plates spread outward slowly, moving perhaps an inch or two each year. Below these plates is a much thicker layer called the mantle, and out in the middle of the Pacific Ocean is a stationary "hot spot" in this mantle, a place where molten rock or magma forces its way upward. When the magma erupts with sufficient volume, it pushes through the plate above it. If the plate remains over the hot spot long enough, successive lava flows eventually build up to sea level and above from depths of several miles. Eruptions that have been of long duration have thus formed the islands of Hawai'i. The piece of the earth's crust that contains these Islands, called the Pacific Plate, is moving slowly to the northwest. As the plate moves, it drags each newly formed island away from the hot spot below. Eruptions continue at the hot spot, and new islands are formed.

This hot spot first became active at least 70 million years ago, when the last of the dinosaurs still roamed the continents. It formed the islands in the Northwest Chain as far back as 40 million years ago. Kaua'i, the oldest of the main islands, emerged from the sea less than 6 million years ago, while the Big Island is still growing as volcanic eruptions fill shallow coastal waters with lava. Thirty miles to the southeast is Lo'ihi Seamount, the next Hawaiian island, which has pushed to within three thousand feet of the ocean surface as lava forces its way from the mantle through the crust above.

The geologic evolution of an island doesn't end with its volcanic formation. As an island glides away from the hot spot it cools and contracts, it slowly sinks under its own weight, and it erodes from the combined forces of tropical rainfall, ocean waves, and wind until nothing is left but a low sandy island or coral atoll. In time even this disappears, leaving a submerged island or seamount. Thus the islands are formed from below the waves and gradually are drawn back again.

What does all this have to do with the birds in Hawai'i? Those disappearing Leeward Islands may have played an important part in collecting the flora and fauna that existed in Hawai'i before the first people arrived. Successful establishment of new species before human intervention must have been extremely rare on these remote islands, occurring perhaps once every seventy thousand years. If the oldest of the main islands, Kaua'i, has existed for 5.6 million years, then we would expect only eighty species of plants and animals to have made the trip and survived. This is a far lower number than the actual species count.

But what if pioneering plants and animals have had ten or twelve times as long to populate these Islands? It is quite possible that the ancestors of some Hawaiian species we know today first arrived before Kaua'i ever rose from the sea, landing on islands of the Northwest Chain when these islands were much younger, larger, more hospitable, and farther to the southeast than they are today. As new islands formed over tens of millions of years, plants and animals, especially birds, must have had a relatively easy time crossing narrow channels to populate these new homes.

Thus, the biota of Hawai'i did not need to be established by pioneers that found their way to these islands in the comparatively short 6 million years or so since Kaua'i was formed. Perhaps dozens of Hawaiian Islands have spread a sort of net across time and space, collecting lost and drifting life in the middle of the Pacific for 60 or 70 million years. The concept of Hawaiian bloodlines that are older than the Islands themselves is a profound one indeed.

Adaptive Radiation

As if the geologic evolution of these islands isn't astonishing enough, a second natural mechanism has played an even larger role in blanketing these islands with life. Charles Darwin observed the same process in the Galápagos Islands, and his recognition of the forces at work led to the most profound revolution in the history of natural science. Today we call the process adaptive radiation: the evolution of different species or forms from a single species of plant or animal in the face of unexploited environmental opportunities. Darwin's description of adaptive radiation in the Galápagos made the finches of those islands famous. If he had sailed to Hawai'i instead, he would have found that the Hawaiian birds called honeycreepers demonstrate the principle even more remarkably.

The condition that has fueled adaptive radiation in Hawai'i is isolation. The islands of Hawai'i are tiny specks of land in the largest expanse of water on earth. Their total area is a little smaller than the state of New Jersey or Riverside County in California. These islands are farther from neighboring land masses than any other island group in the world: from Honolulu it is 2,400 miles to San Francisco, 3,800 miles to Tokyo, and 5,000 to Sydney.

It seems incredible that such tiny and isolated points of land could be teeming with life, yet the first Polynesians arrived to find islands with thousands of species. Here's how it might have happened, using Hawaiian honeycreepers as an example. Fifteen or 20 million years ago a flock of finchlike birds from Asia or perhaps Central America are drawn away from their continental home, perhaps during a powerful storm. A few of them find their way to an island in the middle of the Pacific. This island supports huge colonies of seabirds and some migratory wetland birds, but few if any other land birds. The newly arrived finches will be ancestors to the honeycreepers, but these pioneers are not particularly specialized. They are able to feed on a variety of seeds and perhaps a few insects. These birds are not well adapted to life on their new island home, but in this isolation there is no competition at all for the ample food resources at hand, and there are no predators. They find what they need to survive, and they multiply.

There are other islands near the birds' new home, too distant for constant back-and-forth flight but close enough that individuals or small flocks occasionally make it from one island to another. The birds that make it to a new island are separated from others of their kind and they evolve differently over a long period of time, exploiting different food resources or behaving somewhat differently. They may add flower nectar to their diet. Then some of them recolonize the island from which their ancestors came. This first island now has two bird species that do not share exactly the same food resources and do not interbreed. Over time, new islands form through volcanic action, and the evolving honeycreepers island-hop to colonize them. Through this infrequent back-and-forth island-hopping many new forms, adapted to different living conditions, radiate from common ancestors.

At immeasurably rare intervals some new life form arrives on wind, wings, or waves and establishes itself, tipping the exquisite ecological balance of the Islands. A bird or a bat may offer some competition for food, a new insect may serve as food, a predator such as a hawk or an owl may make food of the creatures that have arrived before it. The incredible isolation of the archipelago ensures that the frequency of these successful colonizations is measured in tens or hundreds of thousands of years.

Eventually, the island that was the first home of the pioneering honeycreepers drifts to the northwest and erodes away until the low, tiny patch of ground can no longer meet the ecological needs of any of the honeycreepers that have evolved, and on this island they die out. But by this time the birds, as well as plants and invertebrates, have colonized many younger, larger islands in the chain. The process occurs very slowly over a very long time. After tens of millions of years, these Islands are brimming with life.

What the First Humans Found

Quite recently in geologic terms, some fifteen hundred years ago, another species made its way to this balanced world of slow changes when Polynesian voyagers first discovered Hawai'i. They did not come alone. If plants and animals had successfully colonized Hawai'i every seventy thousand years, then these people brought a million years of change to the Islands on a few boats. Their domestic animals included pigs, dogs, and junglefowl, with rats, snails, geckos, and skinks probably transported unintentionally. For their gardens they brought bananas, coconuts, sweet potatoes, taro, and other food plants, along with perhaps half a dozen hitchhiking weeds.

The introduction of so many new life forms all at once would have been disruptive enough, but the Polynesians' way of life caused vastly more disruption of Island ecosystems. To plant their crops they burned most of the lowland forests, driving untold bird species upslope to surviving forests or to extinction. They subsisted not only on the animals they had brought with them, but on a smorgasbord of flightless birds and ground-nesting species that probably knew no fear of humans.

How do we know of the radical changes in the birdlife of these Islands that occurred so long ago? We might never have guessed at the lost avian diversity of Hawai'i were it not for the remarkable survival of ancient bird bones at a handful of sites: limestone sinkholes on O'ahu, sand dunes on Moloka'i and Kaua'i, and lava tubes on Maui and the Big Island. We are beginning to learn that the diversity of birdlife in existence when those first Polynesians arrived was greater than anyone would have believed or even imagined.

The bones that have been pieced together paint an incredible picture of avian riches, of dozens of bird species that perished before Captain Cook made these islands known to the outside world. Before any Western naturalist ever described them, they disappeared: nine geese, in-

Red Junglefowl

cluding some huge flightless species, an eagle, two hawks, three flight-less ibises, ten rails (including the smallest rail ever known), four owls, three crows, a honeyeater, and nearly two dozen honeycreepers. These are the birds whose bones we have been lucky enough to find, but there must have been many more.

Other birds that still survive today were apparently much more plen-tiful on the Islands. Species such as the Palila, 'Ō'ū, Maui Parrotbill, and Nēnē had much larger ranges, often extending into lowlands near the sea instead of the uplands and high forests where they are generally restricted today. Many seabirds that now nest only on offshore islets or in very restricted areas on the main islands apparently nested over much wider areas before the advent of the first humans, with their rats and dogs.

By the time Captain Cook arrived in 1778 and brought these Islands to the attention of the world, it is likely that half or more of their endemic bird species had already perished through the intervention of humans and the animals they brought with them.

Recent Arrivals

The native birds of Hawai'i make up only a part of the state's current avian diversity. The Polynesians who discovered Hawai'i began a long history of bringing new birds to the Islands. They brought with them the Red Junglefowl or Moa, chicken of the Polynesian people. Wild Jungle-fowl can still be seen at Kōke'e State Park on Kaua'i. Subsequent immi-grants to the Islands brought a constant stream of alien birdlife. Today nearly every bird habitat in Hawai'i is occupied by some introduced birds. They have been brought for many reasons: in misguided attempts to control introduced insect pests, out of nostalgia to see more of the birds from back home, to be hunted for sport, or as pets that later escaped.

There are birds in Hawai'i that will be familiar to visitors from all corners of the globe. Some of the introduced species bear names that reflect their heritage: Japanese White-eye, Chinese Thrush or Hwamei, Eurasian Skylark, Java Sparrow, Brazilian or Red-crested Cardinal, and Kentucky or Northern Cardinal. And of course, the ubiquitous House Sparrow that seems to shadow humanity everywhere can be found in Hawai'i, too.

The field checklist of birds published by the Hawai'i Audubon Soci-ety lists 165 species that breed in Hawai'i or regularly visit the state. Coincidentally, this may be roughly the number of native bird species that existed when the Polynesians arrived. In a sense, we removed dozens of the birds that evolved in Hawai'i and replaced them with birds of our own choosing. But as colorful and delightful as an introduced Red-crested Cardinal may be, most of us would rather see an endemic 'I'iwi. Hawai'i is a unique and spiritual place, and its endemic birds fit seamlessly into the web of life as no introduced species possibly can. Our challenge for the future will be to protect the unique Hawaiian birds that remain and to keep from further disturbing the delicate balance of Island life. Perhaps, if we are successful, some of the birds we have brought to Hawai'i will evolve into new species that will dwell in the forests of Hawaiian islands yet to emerge from the sea, a few hundred thousand generations from now.

3

New Species and New Challenges

SCIENTISTS CONSIDER HAWAI'I an unparalleled "natural laboratory" where a small number of pioneering species were presented with conditions that enabled them to evolve wildly, filling virtually all the diverse habitats that the Islands offer. When a species with the right basic qualities for survival arrived, it found a laboratory that was stocked with plenty of opportunities and very few limitations. Birds could nest on the ground without fear of predators. Many even lost the ability to fly. Plants had little need for thorns or other protective measures because there were no large herbivores to fend off (except perhaps for huge grazing geese that we know only from their bones).

This grand evolutionary experiment continued for 70 million years, turning out thousands of new species. Then, quite abruptly, this elegant process was radically altered when additional elements were added to the experimental formula. The initial ingredients of isolation, unexploited resources, and benign surroundings were overwhelmed by more potent ingredients such as increased competition, predation, habitat modification, and disease. All were brought into the laboratory, intentionally or otherwise, by one of the last species to find these Islands independently and become established. That species, of course, is *Homo sapiens.*

Like hapless sorcerer's apprentices, we tinkered with the complex apparatus by bringing new plants and animals to the Islands and modifying the habitats we found, sending the experiment off in new and sometimes alarming directions. In the process, we made a mess in the laboratory and destroyed many of the species that were its unique and irreplaceable products. Today, we ask biologists and land managers to stop the reactions that we have set in motion, while we continue to tinker and stir new ingredients into our brew.

Our understanding of Hawaiian birds is incomplete without knowledge of the ecological changes that we have made in Hawai'i and how

24

these changes are affecting Island birdlife. So let's take a peek into the laboratory, identify the plants, animals, and activities that have disrupted the evolutionary experiment, and look at the processes that are taking place today in the "natural laboratory."

Introduced Species

The plants and animals introduced to Hawai'i form a very diverse list. Some have had no apparent effect on endemic birdlife, while others have been devastating. The introduced species that are serious threats include large habitat-destroying mammals, disease organisms, predators, invasive plants, and introduced birds. They affect Island birdlife by competing for food and habitat, altering the habitat that endemic birds evolved to occupy, preying on endemic birds, or infecting them with disease.

Among the introduced species, mammals did the earliest and most visible damage. Goats, sheep, cattle, and pigs all eat native plants and have caused massive disturbance of ecosystems in the process. Goats were brought to the Islands very early. Captain Cook, on his discovery voyage, left some on Ni'ihau in 1778 and some on the Big Island in 1779. By 1850 feral goats were abundant on all the islands, browsing voraciously on native plants, overgrazing, compacting soil, and reproducing prodigiously. By early in the twentieth century the impacts of huge goat populations had reached the crisis stage. Between 1920 and 1970, about seventy thousand of them were removed from Hawai'i Volcanoes National Park on the Big Island with little impact on their population. Since then, fencing and removal of another fifteen thousand has nearly eradicated them from the park. A similar effort occurred at Haleakalā National Park on Maui where hungry goats threatened the unique silversword plant. Control efforts and fencing have been successful there as well. Nevertheless, goats can still be seen roaming some parts of the Islands. A good place to see them is Kōke'e State Park on Kaua'i, where they nimbly traverse the steep mountainsides.

Sheep and cattle arrived in 1791 and 1793, respectively. Both established feral populations and did serious damage through overgrazing. They ranged through mesic forests, mowing down low growth and nibbling off any new tree seedlings so no regeneration occurred. Over the next century, vast areas of forest were thus turned to grassland. Examples of the effect of grazing on forest plant cover can be seen along Wright Road near the 'Ōla'a Tract of Hawai'i Volcanoes National Park on the Big Island, where fences separate domestic cattle from park land.

Today, there are still some feral sheep and cattle on the Big Island. Sheep and goats near Pu'u Lā'au on Mauna Kea have caused great damage to māmane trees there, the principal food source of the endangered Palila. Recent efforts to reduce these feral animals have allowed the māmane trees to make a comeback, bolstering hopes that the Palila may also be able to survive. A few feral cattle also linger on the Big Island, some of them roaming the windward forests in the area of the Hakalau Forest National Wildlife Refuge.

While sheep, cattle, and goats did their damage swiftly, the pigs apparently took their time. Polynesians introduced pigs to Hawai'i long before the other domestic animals arrived, but the diminutive Polynesian pig may have done little damage to native forests. Later, the larger European pigs bred with their smaller Polynesian cousins, and the resulting hybrids felt quite at home in the forest. (This could be related to the introduction of the earthworm, which provided a protein source for forest pigs.) The rampaging descendants of European pigs still do terrible damage to the areas where native birds dwell, uprooting tree ferns to feed on the fleshy core of the trunk, digging up the forest floor to feed on roots and earthworms, destroying vegetation, and causing erosion. They leave wallows in the forest where rainwater collects and mosquitoes can breed. The pigs are secretive and seldom seen, but you needn't go far to see evidence of their presence. There is usually fresh rooting along the 'Aiea loop trail at Keaīwa Heiau State Park on the outskirts of Honolulu. Fencing is the only sure way to protect pristine forest from pigs, and it is a common management strategy at many refuges, sanctuaries, and preserves in the state.

The harm done to Hawaiian birds by grazing and browsing animals is indirect, through modification of habitat. Introduced predators, however, take their toll directly. The principal culprits are rats, cats, and mongooses. First to arrive were the rats, probably as stowaways with the first Polynesians, while the ships of later visitors brought other rat and mouse species to round out the rodent forces. Rats raid the nests of ground-nesting birds, climb trees to the nests of other birds, and compete with birds such as the 'Alalā, 'Ō'ū, and endemic thrushes by eating fruit. Rats also limit the reproduction of some native plants by eating their seeds.

Early sailors controlled rats aboard their vessels with cats, allowing feline immigrants to arrive in Hawai'i with some of the first sailing ships. Once ashore, cats wasted little time before sampling the local cuisine. They are notorious birders all around the world, and the situation in Hawai'i was no different. Cats probably exterminated flightless rails in

Hawai'i and certainly decimated other ground-nesting birds. Today there are populations of feral cats on all the main islands, where they prey on ground-nesting birds and any others they can catch in the trees. Birds of the understory such as 'Elepaio are particularly at risk. Control of cats is vital to the survival of seabirds such as the Red-footed Booby at Kaneohe Marine Corps Base Hawaii on O'ahu and Kīlauea Point on Kaua'i. Cat populations can rise to alarming levels at forest parks near urban areas, such as Wa'ahila Ridge on O'ahu.

If rats were brought here accidentally, and cats redeem themselves as aloof but endearing pets, the small Indian mongoose can offer no such excuses for its existence in Hawai'i. It was introduced in 1883 to reduce rat populations in the sugarcane fields. Conventional wisdom, offered up by tour guides, is that mongooses don't eat rats because the mongooses prowl during the day and the rats are out at night. Actually, the mongoose diet includes some rodents, but rats have teeth, birds have none, and eggs are irresistibly tasty. Consequently, the mongoose is a serious predator of nearly any bird it can reach, including ground-nesting species, ducks, and fledglings that may rest on the ground.

Once mongoose predation is recognized as a threat to a species, human intervention can be helpful. Mongoose and cat control at Hale-

A mongoose

akalā has increased the nesting success of Dark-rumped Petrels that dwell in burrows there. Fortunately for the birds of Kaua'i, the mongoose has not become established on that island.

It isn't just toothy little mammals that prey on birdlife. Even some introduced birds pose predatory threats to endemic species. Barn Owls sometimes take young seabirds, including White Terns in Kapi'olani Park. The Common Myna also preys on the eggs and young of other birds, although Myna populations are generally low in the forests where many endemic birds are found.

Other introduced birds pose a threat by competing with endemic species and harboring avian diseases. Populations of introduced birds are greatest in the lowlands where native plants and animals have been largely wiped out, but many introduced birds are making inroads into the less disturbed habitats. Some aliens that have been particularly successful at invading native forests include Japanese White-eye, Red-billed Leiothrix, Northern Cardinal, Japanese Bush-Warbler, and Kalij Pheasant. Introduced birds act as reservoirs for diseases such as avian malaria and pox. Most introduced birds have substantial immunity to these diseases and act as carriers, exposing endemic birds. Hawaiian species, evolving in the absence of disease, show little or no immunity to these diseases and can be weakened or killed by them.

These avian diseases are usually spread from bird to bird by another group of winged aliens: mosquitoes. As disease vectors, mosquitoes are perhaps as lethal to endemic Hawaiian birds as any of the other introduced animals. Recent research has shown that one bite from a mosquito carrying avian malaria is enough to kill an 'I'iwi. Fortunately, the few species of mosquitoes in Hawai'i are all tropical varieties that are limited to lower elevations. Unfortunately, these tropical mosquitoes are gradually extending their ranges into cooler high elevation areas where endemic Hawaiian birds are restricted, assisted by pigs whose wallows serve as perfect pools for mosquito breeding. It is no coincidence that the dwindling ranges of several surviving endemic birds lie above the altitudes where these tropical mosquitoes have penetrated. Introduction of mosquitoes from cooler parts of the world might bring swift extinction to many remaining endemic birds in Hawai'i.

Today, some of the greatest anxiety among biologists and especially ornithologists is over another animal that threatens to make its way to Hawai'i. The brown tree snake is a Hawaiian ornithologist's worst nightmare, and the nightmare has already become reality on the island of Guam. Sometime around 1940, this reptilian native of New Guinea and the Solomon Islands made its way to Guam, probably as a stowaway in cargo. It is agile and inquisitive, and it likes to get up off the ground where it can feed on birds and their eggs. The result has been a decimation of birds on Guam. The mildly venomous tree snake is blamed for the disappearance of nine of the eleven native species of forest birds on Guam since 1975. The snake's arboreal habits have also taken it atop power poles, where it routinely short-circuits the island's power system. A fertile creature with no enemies, the snake has proliferated to the point where there are as many as twelve thousand per square mile in some parts of Guam. Even if there were any birds left in Guam's forests, would you want to go walking there?

The snake's climbing instincts have prompted it to slither up the landing gear of military planes on Guam: next stop, Honolulu. Several snakes have already been transported in this manner, but apparently all have been dead on arrival or quickly intercepted. Successful invasion by just one fertilized female would almost certainly create havoc for both the birdlife and the visitor industry in Hawai'i.

Even introduced plants can pose a threat to Hawaiian forests and the birds that dwell there. Some aggressive introduced plants crowd out native plant species that birds depend on for food. One of the worst is banana poka, a vine that can grow to the tops of 'ōhi'a and koa trees and smother them. This vine can be seen in many Hawaiian forests. Look for it along the roads at Kōke'e State Park on Kaua'i. Another plant pest is guava. Various species of guava can grow into thick monoculture stands that crowd out other trees. The guavas are spread by pigs that eat the fruit and deposit the seeds elsewhere, complete with a little fertilizer. Even plants such as ornamental ginger and lantana can crowd out the seedlings of native species. Both are widespread—look for ginger in the forests of Hawai'i Volcanoes National Park, and both plants at Kōke'e.

Banana poka

Human Impacts

We have upset the balance of life in Hawai'i not only by introducing alien species, but also through our activities on the Islands. In a few cases our impacts have been direct. The early Hawaiians ate birds they caught, and probably exterminated several species this way, including some large flightless birds. They also killed some birds for their feathers. Later human inhabitants harmed bird populations in similar ways. Feather hunters decimated seabird colonies on the Northwest Hawaiian Islands in the early 1900s for the millinery trade, prompting Teddy Roosevelt to establish the Hawaiian Islands Bird Reservation (later renamed the Hawaiian Islands National Wildlife Refuge) in 1909.

More significant than this direct action against birds, though, has been habitat modification. Some of this modification occurs as a result of introduced animals and plants, as described above. Much of it occurred during Hawai'i's prehistory, when early Hawaiians cleared vast lowland areas for agriculture. Disturbingly, damage to critical habitat continues today. It is fueled by commercial and recreational pursuits that are not always compatible with habitat preservation.

The most obvious modifications to Hawaiian habitats occur through land development—the building of roads, houses, hotels, and the like. Most development occurs in the lowlands where there are few endemic birds left, but there are exceptions. A drive up Kaloko Drive above Kona on the Big Island may leave you shaking your head that any development was permitted in this formerly high-quality endemic bird habitat. I saw my first 'I'iwi in this forest, but it is increasingly difficult to find them on the scarred slope of Hualālai.

Cattle ranching can also conflict with management of habitat for endangered birds, particularly on the Big Island. Grazing by cattle can reduce forests and woodlands to grasslands, and prevent new growth of any seedlings. Forest areas can also be separated by intervening tracts of pasture, fragmenting available habitat. The forests that are well suited for cattle grazing are also an important habitat for 'Alalā, Palila, 'Akiapōlā'au, Hawai'i Creeper, and 'Ākepa. In many of the same areas, logging of valuable koa and 'ōhi'a also disrupts native forests. Essential habitat needs to be protected through purchase and good stewardship, or through incentives for landowners to protect native ecosystems on their property.

How You Can Help

Despite the many pressures facing Hawaiian birds, there are some reasons for optimism. Appreciation for the unique plants and animals of Hawai'i is growing, and there are increasing efforts to protect endemic species. The U.S. government has established a national wildlife refuge for the protection of forest birds at Hakalau on the Big Island. This is the first national wildlife refuge in the nation established for the protection of forest birds. The State of Hawai'i recently established the Pu'u Wa'awa'a Wildlife Sanctuary on the slopes of the Big Island's Hualālai, specifically for 'Alalā habitat. The Nature Conservancy has been expanding its efforts in Hawai'i, protecting vital habitat. The Sierra Club and the Hawai'i and National Audubon Societies are working to protect endemic species and their habitats. The worldwide increase in ecotourism is demonstrating to the Hawaiian visitor industry that there is economic value in endangered plants and animals.

While many of the efforts to protect endemic birds need to be carried out by government agencies or well-financed private organizations, there is still much that individuals can do to make a difference. Whether you are a lifelong resident of Hawai'i or a visitor staying for only a week, there are many actions you can take to help ensure a positive future for Hawai'i's birds and natural places. Here's a list to guide you.

Don't spread plant pests. This was covered in Chapter 1, but it bears repeating. Plants can be transported inadvertently when seeds stick to your clothing and shoes. Take care not to bring alien plants into Hawai'i, or transport plant pests within the state.

Report sightings of animals that don't belong. You probably won't see snakes or alligators anywhere in Hawai'i, mongooses on Kaua'i, or any introduced mammals in preserves or national parks. If you do, report them at once to the State Division of Forestry and Wildlife or other authorities.

Don't put additional pressure on Hawaiian birds. No matter how lightly you tread, or how careful and respectful you are, your mere presence can be a disturbance to endangered birds. If you visit the spots described in this book, you have a good chance of seeing twenty endemic species or subspecies. You won't see some of the rarest birds or visit the most critical habitat. Take your satisfaction from working toward their continued preservation instead.

Remove bird hazards. Discarded plastic six-pack holders and snarled fishing line can tangle or choke birds. Try to leave birding areas cleaner than you found them.

Report sightings of rare birds. If you are lucky enough to see one of the rarest Hawaiian birds, or a bird not known to occur regularly in Hawai'i, be sure to report it to the U.S. Fish and Wildlife Service, the State Division of Forestry and Wildlife, or the Hawai'i Audubon Society.

Support organizations working to protect Hawaiian birds. Public interest organizations that acquire habitat, monitor bird populations, influence and assist government agencies, and educate the public are able to carry out their activities largely through membership donations. Some of these organizations include the Hawai'i and National Audubon Societies, the Sierra Club, The Nature Conservancy, and the Natural Resources Defense Council.

Volunteer your time and effort. Several groups in Hawai'i organize activities such as tree planting, trail restoration, and alien plant removal. Some of these groups are The Nature Conservancy, Sierra Club, Hawaiian Trail and Mountain Club, and the Hawai'i Nature Center. (Additional information on these groups is in the appendix.) You can also help gather valuable data by participating in Hawai'i Audubon Society birdathons or Christmas counts. Hawaiian birders eagerly welcome newcomers.

Stay informed about current issues and speak out. If you are aware of current issues, you will be able to speak out when your voice is needed. Publications of the organizations listed above will help you stay informed. Support legislation and government programs designed to protect Hawaiian birds and their habitats. Write to government agencies responsible for protecting Hawaiian wildlife such as the U.S. Fish and Wildlife Service, the National Park Service, and the Hawai'i Department of Land and Natural Resources, and let them know that you support greater protection for Hawaiian species, including birds.

Let the visitor industry know that wildlife is important. Ask hotels what conservation efforts they support and what interpretive programs they offer before you make your reservations. Avoid developments and forms of recreation that put undue strain on Hawaiian resources. Write to the Hawai'i Visitor Bureau and the Chamber of Commerce. Let them know that Hawai'i's environment is important to tourism.

Tell your friends. Let others know about the unique birds of Hawai'i, the threats these birds face, and the help they need. The more people who are working for the protection of Hawaiian birds and their habitats, the better.

Teach your children. The next generation may be the last with an opportunity to save some endangered species. Share your love and appreciation of nature with your children. Volunteer your time to work on education programs for children through organizations such as the Hawai'i Nature Center and the Hawai'i Audubon Society. If your community doesn't have such a program, help start one.

Think and act positively. The threats that face Hawaiian birds and their habitats are daunting, but not overwhelming. It is within our power to preserve much of the unique life that remains, if every concerned individual contributes to the effort. We must recognize and publicize the aesthetic and economic value of intact forests. We should support development that is good for all of Hawai'i, and reject growth that trades narrow or short-term gain for permanent damage. If we treat these Islands as if we are readying them for our grandchildren and their children, then some of the Hawai'i we know and love will remain for them to enjoy.

4

Oʻahu

AMONG THE HAWAIIAN ISLANDS, Oʻahu is known as "the gathering place," and the name is quite appropriate. Three-quarters of the state's 1.1 million residents live on Oʻahu, and most visitors spend time on the island or at least pass through on their way to other island destinations. Oʻahu is a gathering place for birds, as well. There are more species of introduced birds here than on the other islands, reflecting hundreds of years of immigration and intensive human activity. Christmas bird counts conducted by the Hawaiʻi Audubon Society in Honolulu generally record a total of about fifty species, and half of these are introduced birds.

Oʻahu is the third-largest landmass in the state, behind the Big Island and Maui. Two mountain ranges comprise Oʻahu: the Koʻolau Range that forms the backdrop for Honolulu, and the Waiʻanae Range along the western edge of the island. These mountains are low in comparison to the 10,000-foot-plus peaks of the Big Island and Maui: the highest point in the Koʻolau Range is 3,150 feet, and Oʻahu's tallest peak is 4,020-foot Kaʻala in the Waiʻanae Range.

Sadly, the populations of endemic birds on Oʻahu have declined to very low levels. When the U.S. Fish and Wildlife Service surveyed forest bird populations in the state between 1976 and 1984, they didn't survey the birds on Oʻahu. This was partly because the populations were just too small and widely dispersed for accurate counting. The elevation of Oʻahu mountains provides only weak protection against disease-bearing mosquitoes from the lowlands. Other introduced predators such as rats and mongooses abound, and the cities and towns provide constant sources of feral dogs and cats.

Still, there is reason for hope. A few endemic bird species have survived on Oʻahu even after fifteen hundred years of human habitation, first by early Hawaiians and then by westerners. Today there is spreading concern for endangered species and dwindling habitats, and there are new efforts to protect, preserve, and enhance what is left. With our help, the populations of unique plant and animal species that still occur in viable numbers will survive to remind us of the true natural history of this place.

In the Ko'olau Range, a few endemic bird species are still found despite urban encroachment from nearly all sides. Their population densities are rather low, but with luck you should be able to see O'ahu 'Amakihi and perhaps 'Apapane and 'Elepaio. The Wai'anae Range, higher and more remote from population centers, also supports populations of endemics. Unfortunately, there is very little public access to good birding areas in these mountains. Occasionally the Hawai'i Audu-

Map 1
O'ahu Bird Sites

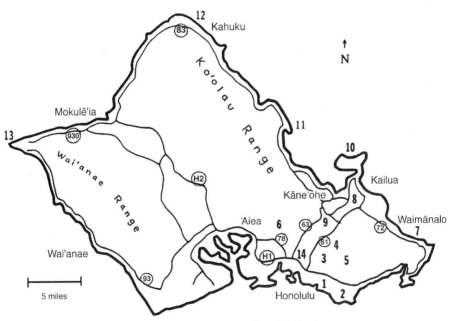

1. Kapi'olani Park
2. Diamond Head
3. Makiki Valley
4. Lyon Arboretum
5. Wa'ahila Ridge
6. 'Aiea Loop Trail
7. Makapu'u Point—Waimānalo Coast

8. Kawainui Marsh
9. Ho'omaluhia County Nature Park
10. Kaneohe Marine Corps Base Hawaii
11. Kualoa Regional Park
12. Kahuku Area Wetlands
13. Ka'ena Point
14. Rainy Day Birding—Bishop Museum

bon Society gets permission to enter private property in the Wai'anae Range, and The Nature Conservancy offers a couple of hikes each month into its Honouliuli Preserve. The beaches of O'ahu are good for observing marine birds because there are many small offshore islets that provide safe nesting areas.

The long coexistence of birds and people has fashioned O'ahu into an island where birds can be found in the most improbable places. White Terns, common on islands of the Northwestern Chain but uncommon on the main islands, nest in Kapi'olani Park within a stone's throw of high-rise hotels. The endemic O'ahu 'Amakihi visits the gardens of Honolulu residents at the foot of the Ko'olau mountains, and can often be seen at a local park in the neighborhood. Across the island, a huge colony of Red-footed Boobies exists alongside a Marine Corps rifle range.

The visiting birder will find that most of the lodging choices on O'ahu are in the Waikīkī area of Honolulu. This densely populated neighborhood is actually a good home base for birders: nearby Kapi'olani Park offers a profusion of introduced birds, while some of the best spots to see forest birds and marine birds are within a few miles.

Most of the people in Hawai'i live on O'ahu, so most organized birding and hiking activities are conducted on this island as well. The Hawai'i Audubon Society leads monthly outings that often visit sites not open to the public. Several other organizations lead hikes that may be of interest to birders. See the appendix for more information.

1. Kapi'olani Park

FEATURES:
- Great variety of introduced birds
- Very convenient for visitors

FACILITIES:

Just a short walk from most Waikīkī hotels is Queen Kapi'olani Regional Park. This 140-acre park, donated to the people of Honolulu by Hawaiian King Kalākaua in 1877 and named after his wife, is a beehive of activity. It is home to the Honolulu Zoo and the Waikīkī Aquarium (entry fees required), a bandstand, driving range, archery range, jogging path, and amateur sports fields. The park is also a good spot to see many of the common introduced birds on O'ahu. Most of the birds will be easy to identify, but birding in this urban park may test your identification skills: many escaped or released cage birds can turn up, as well.

A highlight of the park is its population of White Terns. First observed on O'ahu in 1961, these birds may have colonized the area because all available nesting sites are occupied in the Northwestern Hawaiian Islands, where they are more common. At first glance you may mistake these birds for the far more common white Rock Doves (Pigeons) in the park. However, the terns are easy to distinguish by their more slender bodies and wings, their prominent black eyes, and their long, straight, black bills. Most of the terns nest in banyan trees, but they are easier to spot when they nest in ironwood trees. Look for them in the ironwood grove at the Diamond Head end of the park. In some years, over thirty pairs nest in the park. Nesting occurs almost year-round, with the greatest activity in the spring and very little nesting from September through November.

Common introduced birds in the park include the Zebra Dove, the larger and less numerous Spotted Dove, and the House Sparrow, House Finch, Red-vented Bulbul, Japanese White-eye, and Red-crested Cardinal. The Java Sparrow seems to be increasing in the area (as it is almost every-where) and is often seen in open grassy areas feeding with Nutmeg Man-nikins. The Nutmeg Mannikin is also known as Ricebird or Spotted Munia. Less common but often found is the Northern Cardinal. Patient birders can usually see Yellow-fronted Canaries feeding in the ironwood trees. From August through April, the native Kōlea is quite common.

A bird that is abundant at the park, and in many lowland areas of all the islands, is the Common Myna. People familiar with the European Starling may recognize the Common Myna as a member of the same family. The Myna was introduced to Hawai'i in 1865 from India, in the hope that it would reduce the army worm infestation in pasture grass. The bird has prospered and now numbers in the millions. The Myna is a little more wary than other urban birds such as Zebra Doves and House Sparrows, but it is always easy to observe in urban areas.

Mynas are communal roosters, which means that when they are not nesting they descend by the hundreds or even thousands upon the tree of their choice at dusk to chatter and talk before settling down for the night. This is an amazing and very noisy spectacle; just don't stand directly beneath the occupied tree. The birds move their roosts from time to time, so I can't direct you to one; ask around to find one near you.

In the spring some birds leave the roost and go off by themselves to nest, and in the process they reveal a rather odd affinity: Common Mynas are afflicted with a cellophane fetish. They find the material irresistible. They'll dodge cars and seemingly risk their lives just to retrieve a cigarette wrapper from a busy street. A study of twenty-five Myna nests found that every one included at least some cellophane in the nest construction.

One of the best birding areas at Kapi'olani Park is the Honolulu Zoo. The extensive collection of exotic and native birds, together with all the plants, animal food, and insects to be found in the zoo, attracts quite a variety of free-flying birds. The African Savannah display may be the best spot in the park to see Common Waxbills. The zoo is open daily from 9:00 A.M. to 4:30 P.M. except Christmas and New Year's Day, there is a small admission fee, and access for the disabled is excellent.

In addition to the common birds mentioned, keep an eye out for avian surprises in Kapi'olani Park. Some uncommon bird species to watch for are Rose-ringed Parakeet, Red-crowned Amazon Parrot, and Lavender Waxbill. Other new species are bound to appear as well. For example, a flock of at least eighteen Red-masked Conures was spotted in the park during the 1994 Audubon Christmas Count. This bird had been spotted a few times before but was not known to be established in Hawai'i, and it does not appear on checklists of Hawaiian birds. Look for this thirteen-inch green parrot with a red face behind the tennis courts and archery range on Pākī Avenue, near the Diamond Head end of the park.

2. Diamond Head

FEATURES:

• Hiking opportunity close to Waikīkī
• A view of this landmark from the inside
• Chance to see some uncommon alien birds

FACILITIES:

The silhouette of Diamond Head is a scene that is symbolic of Hawai'i and especially of Waikīkī. Many visitors to Hawai'i are surprised to learn that Diamond Head is not a mountain but an extinct volcanic crater with a small military base on the crater floor. For visiting birders and hikers, the trail leading up the inside wall of the crater offers a convenient recreational opportunity just a short distance from Waikīkī.

The trail to the top of Diamond Head, or Lē'ahi as it was called by the Hawaiians, begins on the crater floor. The road to the trailhead and military installation tunnels through the crater wall. To get there, take Monsarrat Avenue or Diamond Head Road around to the back or inland (mauka) side of the crater, where the well-marked road passes through the entrance gate and mountain to the trailhead inside. The inside of the crater is open to the public from 6:00 A.M. to 6:00 P.M.

You may not be rewarded with any rare bird sightings on the hike to the top of Diamond Head, but the views of the coastline and Honolulu from the abandoned military fortifications at the summit are well worth the climb. The trail to the summit is easy to follow. The first third is paved and gently sloping, passing through the kiawe forest that covers the crater floor and lower walls. This part of the trail has good wheelchair access. Next is a rocky, unpaved section with a pipe handrail. Finally, you climb hundreds of concrete steps and pass through unlighted tunnels and stairwells to reach the top. A flashlight is ABSOLUTELY REQUIRED to get through these dark structures. Total one-way distance of the trail is 0.7 miles, and it climbs 550 feet in elevation.

Along the way, you will see many of the common urban birds that can be found elsewhere in Honolulu. In this natural setting they all seem to act more like wild birds and less like tame park creatures. Species to look for include Common Myna, Zebra Dove, Northern Cardinal, Red-crested Cardinal, Red-vented Bulbul, Japanese White-eye, Northern Mockingbird, House Sparrow, House Finch, Nutmeg Mannikin, and Java Sparrow.

This spot is highly recommended for families with children. The hike is not too difficult, and the tunnels and fortifications on the mountain are the ultimate kids' fantasy. Plan to come early to avoid the heat and the crowds on this sunny and popular hike. There is a shady picnic area, water, and rest rooms with good access for the disabled at the trailhead.

The outer slopes of Diamond Head used to be a favorite Honolulu birding area. In the 1960s, populations of several Estrildid finches appeared in the area, including Lavender Waxbill and Orange-cheeked Waxbill. A trail stretched one-third of the way around the mountain's circumference, passing through the Nalā'au Arboretum. Today, the situation is much different. The Orange-cheeked Waxbills have apparently died out, and the Lavender Waxbills are rare, although Common Waxbills are becoming more numerous in the area. The trail is no longer open to the public, and the abandoned arboretum is filled with weeds and trash. At least for now, the Nalā'au Trail can be crossed off the list of good birding spots on O'ahu.

3. Makiki Valley

FEATURES:
- Endemic O'ahu 'Amakihi
- Many introduced birds
- Lush forest setting

FACILITIES:

Honolulu is bordered on the north by the Ko'olau Range, whose jagged peaks catch moisture-laden clouds carried by the trade winds. The forests that thrive on this rainfall are home to many of the common introduced birds of the city, additional alien birds that prefer the forest, and even the endemic O'ahu 'Amakihi, 'Apapane, and 'Elepaio.

Many parks and hiking trails provide access to developed and natural areas in the mountains. The closest ones to downtown Honolulu are in the Makiki/Tantalus State Recreation Area, which is surrounded on three sides by urban development. Nevertheless, this area offers forest hiking trails, solitude, and good birding. A scenic loop road winds through the area, providing roadside birding and good viewpoints of the city and ocean below. This road also provides easy access to good birding sites.

One of the best hiking trails in the area, especially for the visiting birder, is the Makiki Valley loop trail. It is easy to find, not too strenuous, offers excellent birding, and forms a loop so there is no backtracking. Free trail maps are even available at the trailhead. There is also good birding from the paved entrance road leading to the trailhead, where varieties of introduced birds—as well as the O'ahu 'Amakihi—are frequently seen.

To reach the Makiki Valley loop trailhead, take Makiki Street in downtown Honolulu to Makiki Heights Drive. Proceed about one-half mile to a point where Makiki Heights Drive makes a hairpin turn to the left and a long narrow driveway goes straight. Drive through the gate and up this driveway to the State Nursery and Arboretum in Makiki Valley.

This lush valley is also home to the Hawai'i Nature Center (HNC), a nonprofit organization that promotes wise stewardship of the Hawaiian Islands through interpretive programs for schoolchildren and weekend hikes and classes for children and adults. The HNC has free maps of the trails in the area and leads walks in this or other areas of O'ahu weekly. Stop in to see their interpretive exhibits and get a map, but please don't disturb classes in progress.

The loop trail starts at the end of the road. The beginning of the trail can be muddy and slippery if it has been raining, and this moisture provides perfect conditions for mosquitoes, so be sure to bring your insect repellent! The route is actually a triangle with sides formed by three different trails. The most gradual climb is made by starting off to the left on the Kanealole Trail, clearly marked with a sign just past the nature center. After climbing gradually for about three-quarters of a mile, you will come to a fork in the trail. The trail to the right, not identified by a sign, is the Makiki Valley Trail. Follow it for about half a mile, crossing several small streams. Soon you will come to an intersection of several trails identified by signs (a rarity on O'ahu). Follow the Maunalaha Trail back down to the HNC. Total loop distance is about two miles, with a seven hundred-foot gain in elevation.

Bird species you are likely to see include Zebra Dove, Spotted Dove, Northern Cardinal, Red-crested Cardinal, Red-vented Bulbul, Red-whiskered Bulbul, Japanese White-eye, White-rumped Shama, Common Myna, House Finch, House Sparrow, Java Sparrow, Nutmeg Mannikin, and if you are lucky, O'ahu 'Amakihi. In winter, Kōlea can be seen in the lawn areas. A bird that is more likely to be heard than seen is the Japanese Bush-Warbler, although from September to December it is almost silent. Occasionally parrot species are seen in the area, most commonly the Rose-ringed Parakeet. Two uncommon birds whose numbers appear to be increasing are Common Waxbill and Red Avadavat.

The presence of voracious mosquitoes and O'ahu 'Amakihi in the same place is a very good sign. Mosquito-borne avian malaria is deadly to most Hawaiian birds, which have no immunity to it. The 'Amakihi appears to have developed or retained some immunity, helping to protect it from the extinction that has befallen so many other Hawaiian forest bird species.

Along the trail you will see many introduced trees and vines, including at least three edible fruits: guavas, strawberry guavas, and mountain apples. The native koa tree also grows in this area.

The Nature Center, rest rooms, and drinking fountain are all accessible to the disabled. The valley area is open from 7:00 A.M. to 6:30 P.M., but closed on holidays. Nature Center hours are 8:00 A.M. to 4:30 P.M.

4. Lyon Arboretum

FEATURES:

• Lush botanical garden
• Chance to see uncommon aliens

FACILITIES:

In the upper Mānoa Valley, where Honolulu backs right up to the mountains, is the lush Lyon Arboretum owned by the University of Hawai'i. This 124-acre tropical garden is an excellent place to stroll among labeled plant specimens and watch for urban and forest birds.

It rains a lot on these mountain slopes, so the trails can be wet and a little muddy. Concrete stepping-stones help you through the softest places. All the water makes it a haven for mosquitoes, so mosquito repellent is STRONGLY RECOMMENDED. (I'm still scratching as I write this!) The rain can be damaging to your camera and binoculars, so don't forget to bring a plastic bag to protect them.

If these precautions make the place sound dismally wet, muddy, and mosquito infested, do not be discouraged. If you come prepared, this is a wonderful place to observe birds and tropical gardens. It is not heavily visited, so you may have a grassy tropical glade all to yourself. If you whistle back at a singing White-rumped Shama, he may carry on a conversation with you for ten minutes. If a shower is passing over, duck into the rain shelter on the hillside, or take a few moments to visit the gift shop. Most of the items are botanical rather than ornithological, but the volunteers who staff the shop love to chat.

To reach the arboretum, take Punahou Street from central Honolulu. This turns into Mānoa Road; follow it all the way up through the residential area of Mānoa Valley—bearing left at the fork with East Mānoa Road —to the end of the road. A street map of Honolulu may help you find your way.

Birds you are likely to see in the arboretum include the Northern Cardinal, Red-crested Cardinal, Red-vented Bulbul and its shyer, more diminutive cousin the Red-whiskered Bulbul, Spotted Dove, Japanese White-eye, Common Myna, House Finch, and House Sparrow. Java Sparrows and Nutmeg Mannikins can be seen browsing in the grass, and White-rumped Shamas are often heard singing throughout the area. Watch for Kōlea during the months when they winter in Hawai'i. Occasionally, endemic forest birds are seen in the arboretum as well. With luck you might see O'ahu 'Amakihi and perhaps 'Apapane. This area is famous for harboring some uncommon and very interesting alien birds, including the Sulphur-crested Cockatoo and Rose-ringed Parakeet. They're never a sure thing, but watch for them.

The arboretum is open from 9:00 A.M. to 4:00 P.M. Monday through Friday and 9:00 to 3:00 on Saturday; it is closed Sundays and holidays. A small donation is requested. Access is poor for the disabled, as none of the trails are paved.

5. Wa'ahila Ridge

FEATURES:
• Endemic birds
• Native koa and 'ōhi'a trees
• Exhilarating views

FACILITIES:

Wa'ahila Ridge State Recreation Area is right on the mountainous edge of Honolulu, yet this park offers forest hiking trails with stunning vistas, native forests of koa and 'ōhi'a trees, and a good chance to see some endemic bird species.

Koa and 'ōhi'a are two of the most prominent and familiar trees of

Hawaiian forests. Both are vitally important to endemic forest birds throughout the state. Honeycreepers such as the 'Apapane and O'ahu 'Amakihi sip nectar from the red lehua blossoms of the 'ōhi'a. The bark of the koa harbors insects that are important to the diet of several endemic bird species. Both of these trees can be seen along the upper part of Wa'ahila Ridge.

To reach the park, take Kapi'olani Boulevard east (Diamond Head direction) until it passes under the H-1 Freeway and becomes Wai'alae Avenue. The first street on your left will be St. Louis Drive. Follow St. Louis Drive up the hill more than a mile, nearly to the top. Turn right on Peter Street and then left on Ruth Place, which ends at the park entrance. Follow the park road all the way up to the end. (The O'ahu map published by University of Hawai'i Press shows this route very well.)

The lower part of the park is planted with introduced Norfolk Island pine, giving it a very woodsy atmosphere. The trailhead, in this grove of Norfolk pines, is well marked. The route up Wa'ahila Ridge is quite steep in some places, and rocky in others. There are a few spots where you will have to scramble over steep rocks to proceed. Also, since the trail follows the spine of the ridge, there are places where the narrow path is flanked by steep drops on both sides. Views of Honolulu are magnificent.

The trail climbs about seven hundred feet from the trailhead to the end, two miles up the ridge. It passes through introduced forests into an area of many native plants, including koa and 'ōhi'a. You are likely to see O'ahu 'Amakihi probing the lehua blossoms of the 'ōhi'a trees. With luck, you may see 'Apapane flying overhead or in the treetops below you on the flanks of the ridge. You will almost certainly be treated to the melodious and variable songs of the White-rumped Shama, especially during its breeding season from February through July. These birds usually stay hidden in the understory, but they are inquisitive and can be drawn out if you return their song. Watch for Common Waxbills in the tall grass along the trail.

Around the parking area and in the introduced trees at the beginning of the trail, watch for Zebra Doves, Spotted Doves, Northern Cardinals, Red-vented Bulbuls, Red-whiskered Bulbuls, House Finches, House Sparrows, Japanese White-eyes, and Nutmeg Mannikins.

The favorite phrase of Hawai'i weathercasters is "windward and mauka showers." This means that even when it is clear, sunny, and warm down in Honolulu it can be cool, windy, and rainy up in the mountains. Be sure to bring protection from rain.

Park hours are 7:00 A.M. to 7:45 P.M. from April 1 to Labor Day, and 7:00 A.M. to 6:45 P.M. the rest of the year. Admission is free. Only the parking area has access for the disabled; even the rest rooms are up a very steep, paved path.

6. 'Aiea Loop Trail

FEATURES:
• Scenic loop trail
• Some endemic birds
• Site of *heiau*

FACILITIES:

A good spot for seeing introduced forest birds, and maybe an 'Apapane or O'ahu 'Amakihi, is at Keaīwa Heiau State Recreation Area. This state park is in 'Aiea, near Pearl Harbor. A good loop trail through the park offers some great vistas of the harbor and the interior of O'ahu, as well as fine birding. The centerpiece of the park is the ruins of a *heiau*, or temple, used by early Hawaiian healers. Many medicinal plants used by Hawaiians are labeled in the garden area. Note the offerings of flowers, and stones wrapped in ti leaves. This is a spiritual place that is still in use.

To reach the park from Honolulu, take the H-1 Freeway west ('Ewa) until it splits, forming H-1 and Route 78. Stay on Route 78, past the "Hālawa/Stadium" exit, and take the "Stadium/Aiea" exit. The off-ramp becomes Moanalua Road. Follow it a short distance toward the bottom of the hill, and turn right on 'Aiea Heights Drive. Follow this road all the way up the hill to the park at the end of the road.

Take the one-way loop road in the park all the way to the top of the hill, where there is a sign pointing out the trailhead for the 'Aiea Loop Trail. This trail is very well maintained and easy to follow. There are side trails that branch off in a few spots, but the main trail is always quite prominent.

The trail follows a long oval, climbing up the spine of one ridge and doubling back along the top of another. The total length is 4.8 miles, with an elevation gain of seven hundred feet. The trail can be slippery when wet, so be careful. Most of the introduced birds will be seen at the beginning and end of the loop. Near the middle of the loop, at its highest elevation, you will see some native trees, 'ōhi'a and koa, and perhaps some native birds as well.

Common introduced birds in the area include the Japanese White-eye, White-rumped Shama, Japanese Bush-Warbler, Red-crested Cardinal, Northern Cardinal, Common Myna, Zebra Dove, Spotted Dove, Red-vented Bulbul, Common Waxbill, and feral Chicken. If you are lucky you may glimpse an endemic 'Apapane or O'ahu 'Amakihi.

Another endemic bird to look for in this park is the 'Elepaio. Separate subspecies of 'Elepaio are found on Kaua'i, the Big Island, and O'ahu. The birds on the neighbor islands are still fairly common, but on O'ahu the 'Elepaio has declined to the point that it is being considered for listing as a threatened or endangered species. These birds used to be seen regularly along this trail, but there have been very few recent sightings. Look for them in the understory, often just about eye level, where they may approach quite closely.

About one-third of the way around the loop, where it reverses course and heads back, is the unmarked beginning of the 'Aiea Ridge Trail. This trail is narrower and more precipitous than the loop trail, and long pants will be needed unless the trail has been cleared very recently. Right at the beginning of the ridge trail there is also a short dead-end trail. The ridge trail keeps to the left as it takes off from the loop trail. Just a few yards up the trail the scenery will change, as you leave the area of introduced eucalyptus and enter an area of native koa and 'ōhi'a stands. You will almost certainly have the trail to yourself, and the isolation and grandeur of this windy ridge are magnificent. Your best chance to see an 'Apapane, O'ahu 'Amakihi, or even an 'Elepaio in the park is up on this ridge. Even at this remote area, these birds are becoming less common, so patience and luck will dictate what you see.

After you backtrack to the loop trail, continue on the loop to your left. If you see some areas along the trail where it looks like someone has been digging for worms, you're right. Wild pigs often root around for worms, a favorite food. The resulting disturbance can cause trail erosion and destruction of plant life, and can leave a hole that becomes a mosquito breeding pool after the next rain. The inch-long yellow fruits that litter the trail toward the end are lemon guava. Most of the fruit is too high to pick, and the bruised fruits on the ground usually contain tiny white fruit fly larvae, but with luck you may find some edible fruit to taste.

Near the end of the trail, it crosses 'Aiea Stream and comes out at the camping area. Follow the loop road a short distance up the hill to the trailhead where you started.

The park is open from 7:00 A.M. to 7:45 P.M. between April 1 and Labor Day, 7:00 A.M. to 6:45 P.M. at other times. The rugged trail is not accessible to the disabled.

7. Makapu'u Point—Waimānalo Coast

FEATURES:
• Marine birds

FACILITIES:

There are relatively few places along the coasts of the main Hawaiian Islands where marine birds can be observed. Fortunately for birders, the windward coast of O'ahu from Makapu'u Point to Waimānalo Beach offers many opportunities to observe marine birds just a short distance from Honolulu.

Birds are attracted to this area because of Kāohikaipu and Mānana Islands, 0.4 and 0.7 miles offshore, respectively. These islets afford protection from many predators such as rats, cats, dogs, and mongooses that have all but eliminated seabird colonies on the main islands. Mānana Island is the more important of the two islands. Marine birds that breed there include 60,000 to 90,000 pairs of Sooty Terns, 10,000 to 20,000 pairs of Wedge-tailed Shearwaters, 15,000 to 20,000 pairs of Brown Noddies, and fewer than a dozen pairs each of Bulwer's Petrels and Red-tailed Tropicbirds. Black Noddies also breed on Mānana, but information is insufficient to estimate the population. The table below shows when these birds are present on Mānana. Several other species visit the islet or surrounding waters, including White-tailed Tropicbird, Red-footed Booby, Brown Booby, Great Frigatebird, and White Tern. Mānana Island is a sanctuary that can be entered only by special permission, because human footfalls can easily collapse nest burrows and cause massive disruption of colonies.

There have been efforts to encourage the Laysan Albatross to nest on

Table 4
Seasonal Occurrence of Breeding Birds at Mānana Island

BIRD	J	F	M	A	M	J	J	A	S	O	N	D
Wedge-tailed Shearwater			─	─	─	─	─	─	─	─		
Bulwer's Petrel				─	─	─	─	─	─			
Sooty Tern			─	─	─	─	─	─	─	─		
Brown Noddy			─	─	─	─	─	─	─	─		
Red-tailed Tropicbird			─	─	─	─	─	─	─	─		

Compiled from observations, interviews, and information in Richardson and Fisher, "Birds of Islands off Oahu," *The Auk,* Vol. 67, No. 3, July 1950, *Hawai'i Wildlife Plan,* State of Hawai'i, 1983, and *Seabirds of Hawaii* by Harrison, 1990.

Kāohikaipu Islet. These birds are colonial nesters, so biologists try to lure them by creating the illusion of a colony, complete with decoys, ceramic eggs, and recorded calls. If these efforts are successful, albatrosses may become a familiar sight along this stretch of coastline in the future.

Several spots along the coast provide viewing. To reach any of them, head east from Honolulu on the H-1 Freeway, which becomes Highway 72 or the Kalaniana'ole Highway. The road passes Hanauma Bay with its State Underwater Park, then a rocky stretch of coastline, and finally Sandy Beach before turning inland. It next passes the Hawai'i Kai golf course, on the inland side of the road. Just 0.7 miles farther is the trailhead for Makapu'u Lighthouse, a good birding area. A sign at the trailhead reads "Makapuu State Wayside." The route to the lighthouse once served as its access road, so it is paved (but narrow) all the way. To keep vehicles out, there is one locked gate at the trailhead and another one part way up the trail that you must climb over or go around. The old road climbs gradually for just over a mile. The first half of the trail is on the inland side of Makapu'u Head, while the second part offers better views of the ocean. Along the way, look for introduced birds that frequent dry open country, including Zebra Dove, Northern Mockingbird, Red-vented Bulbul, House Finch, Japanese White-eye, Rock Dove, and perhaps Northern Cardinal. The trail ends on the hill above the lighthouse. Most seabirds will be far below you or far from shore, but the view is superb. Be sure to take water and sunscreen, since there is no shade along the trail.

About 0.3 miles beyond the lighthouse trailhead, just where the highway approaches the ocean again, is a small turnout high above the ocean. It commands an excellent view of Makapu'u Point, the coastline to the north, and the island seabird sanctuaries. From this turnout you may also be able to look down at seabirds flying over the water. You can get a little closer to the islets by driving down to Makapu'u Beach County Park, although the seabirds swarming around the islets will still be fairly distant. A spotting scope would be helpful. The status and future accessibility of this park are in question. Some native Hawaiians believe that Makapu'u and neighboring Kaupō Beach Parks are not on public land but on Hawaiian homelands.

Just across the highway from Makapu'u Park is one of O'ahu's most popular tourist attractions: Sea Life Park. Most people come here for the performing dolphins and false killer whales, but this is also a good place to get a close look at marine birds normally restricted to offshore islands or relatively isolated places. Sea Life Park takes in orphaned and injured seabirds such as Red-footed Booby, Brown Booby, Masked Booby, Laysan Albatross, Wedge-tailed Shearwater, and Great Frigatebird. Few visitors spend much time at the bird sanctuary in the park, so you'll be able to step right up and listen to a minilecture on the birds and view any that are receiving care or just hanging out. The facility has booby chicks virtually every year; April and May are the best times to see them. (The park also has conservation programs involving endangered Hawaiian monk seals and endangered green sea turtles.)

Other birds that frequent Sea Life Park are Common Myna, Zebra Dove, Red-crested Cardinal, Red-vented Bulbul, and House Sparrow.

The park is open every day of the year, from 9:30 A.M. to 5:00 P.M. An admission fee is charged. Access for the disabled is excellent.

The last birding stop along this stretch of highway is Waimānalo Beach County Park. Lawn, picnic tables, and trees make it the most hospitable of the county parks in the area. It is a little farther from the island bird sanctuaries, but there are other rewards. The sandy beach attracts winter migrants including Wandering Tattler, Sanderling, and Ruddy Turnstone. Kōlea are common on the lawns during winter. Red-footed Boobies can sometimes be seen at very close range over the beach or surf, flying down the coastline. Great Frigatebirds are commonly seen soaring overhead. Other birds to look for include Common Myna, Zebra Dove, Spotted Dove, House Sparrow, House Finch, and Red-vented Bulbul.

Disabled visitors will find that the rest rooms are accessible, and a short, paved path allows beach, sea, and islet viewing. Waimānalo Beach Park is open at all times, with no admission charge.

8. Kawainui Marsh

FEATURES:

• Wetland birds

FACILITIES:

 (at Kaha County Park)

One of the most significant wetlands in Hawai'i is Kawainui Marsh, on the windward side of the Ko'olau Range, just a few minutes from Honolulu on the Pali Highway, Route 61. This marsh covers several square miles, quite large compared to most other wetlands in the Islands. The area has undergone many changes in the centuries since Polynesians first arrived, and more changes are certain to occur in the future as the sometimes conflicting demands for wildlife habitat, flood control, agriculture, urban development, and water are balanced.

When the first Hawaiians came to this spot, they probably found a saltwater lagoon between the mountains and the sea. They used the abundant runoff from the towering mountains to irrigate taro patches, and erosion from their activities accelerated the natural filling of the lagoon with silt and organic matter. Eventually this silt nearly sealed the lagoon off from the sea, and freshwater replaced salt water, transforming the area into a freshwater marsh about five hundred years ago.

People who lived in the area in the 1920s to the 1940s recall a marsh with a lot of open water, where thousands of ducks would gather in the winter. Vegetation has gradually filled much of the open-water area and reduced its value as a wildlife habitat. Duck populations have declined sharply here, as they have everywhere in the Islands.

Development around the marsh has gradually increased the runoff that flows into it, and local floods have become progressively worse over the years. In the early morning darkness of New Year's Day 1988, the twenty-three inches of rain that had fallen in the preceding twenty-four hours could no longer be contained by the marsh. Floodwaters poured into residential areas and flooded hundreds of homes. The federal government began planning additional flood-control measures that, when constructed, will certainly have an effect on the marsh.

Changes to the marsh will also affect wetland life, including the four endangered Hawaiian wetland birds that breed at Kawainui: the Black-necked Stilt, Hawaiian Coot, Common Moorhen, and Koloa. Today's sensitivity to environmental concerns offers some hope that changes to the marsh will benefit these wetland birds as well as the other residents of Kailua. Additional wildlife viewing areas will probably be constructed, and a visitor center is planned for the marsh area.

Until these facilities are available, the wildlife viewing at Kawainui will be limited. Today the best viewing of wetland birds is not at the marsh itself, but along nearby Ka'elepulu Stream. The stream flows east from the marsh through Kailua town. To reach the area from Honolulu, take the Pali Highway (Route 61) over the mountains and down into Kailua. Turn right on Hāmākua Road, and park near the point where the stream borders the road. There is no parking on the street, but ample parking is available at the nearby shopping center.

A sidewalk makes for easy access to the edge of this marshy stream. You may see Cattle Egret, Black-crowned Night-Heron, Kōlea, Black-necked Stilt, Common Moorhen, Hawaiian Coot, Koloa, and Mallard. In winter a few Northern Shovelers may be present. Walk up the road to the point where it crosses over the stream. There are hundreds of tilapia fish in the stream that will rise to take bread crumbs (or almost anything else) dropped into the water.

The streamside area was formerly private property used for cattle grazing. It has been donated to Ducks Unlimited, a nonprofit conservation organization dedicated to protection of waterfowl and wetlands. This group plans to restore the marsh and donate it to the State. Plans include removal of invasive weeds, construction of secure nesting areas, and an educational kiosk.

There are a few other access points to Kawainui Marsh. At the northeast edge of the marsh is Kaha Park, a grassy playing field that offers views of most of the marsh. Between the park and the marsh there is a canal and a levee. The levee road, which starts at the north end of the park, provides better views of wetland birds. Many people walk and jog along the publicly owned levee even though a sign prohibits entry without permission. Be sure to call the City and County of Honolulu to get permission before you visit. Kaha Park is located at the end of Kaha Street in a residential area of Kailua. It's somewhat hard to find without a street map; the University of Hawai'i map of O'ahu shows the park but does not list its name. The only facilities at Kaha Park are chemical toilets.

9. Ho'omaluhia County Nature Park

FEATURES:

• Variety of introduced birds
• Great views of Ko'olau Range

FACILITIES:

Ho'omaluhia Park is a botanical garden occupying four hundred acres in the town of Kāne'ohe, on the windward side of O'ahu. This park is operated by the City and County of Honolulu, more as a natural forest preserve than a formal botanical garden. Sections of the park have plant specimens from various tropical areas of the world. There are two ponds, several streams, some grassy areas for picnics, and miles of trails that provide quiet, uncrowded places to do some birding. In fact, the name Ho'omaluhia means "place of peace and tranquility."

To reach the park from Honolulu, take the Pali Highway (Route 61) over the Ko'olau Range to the Kamehameha Highway (Route 83). Turn left and go about 2.4 miles to Luluku Road. Turn left again and follow the road to the park. Be sure to stop at the Visitors Center for a trail map.

Nearly all the birds in the park are introduced species, including Common Myna, Zebra Dove, Spotted Dove, Northern Cardinal, Red-crested Cardinal, Japanese White-eye, Cattle Egret, Nutmeg Mannikin, Chestnut Mannikin, Java Sparrow, White-rumped Shama, Red-vented Bulbul, Common Waxbill, and House Sparrow. From August to April the Kōlea is common in the lawn areas. Native Black-crowned Night-Herons and Hawaiian Coots may be seen near the ponds and streams. Occasionally the endemic Koloa is seen. There are also domestic ducks and geese in the park.

In addition to the birds, there are several other introduced animal species that you may spot, including mongooses, turtles, bullfrogs, giant African snails, and huge catfish that look for handouts from visitors.

Frequent nature and bird hikes are offered, and camping is allowed with advance permission. General nature hikes are offered each Saturday at 10:00 A.M. and Sunday at 1:00 P.M., with reservations required. General hikes are also offered on all State holidays at 10:00 A.M. The guides are always happy to point out and identify birds along the trail. Special birding hikes are sometimes scheduled in the springtime. For further information or to reserve space on a hike, call the visitor center (see the appendix). The park is open every day except Christmas and New Year's Day from 9:00 A.M. to 4:00 P.M. Admission is free.

The Ko'olau Range towers over the park and catches clouds, producing rain that sometimes makes the trails muddy and the streams hard to ford. Boots and rain gear are recommended. Mosquitoes can be thick in this moist area, so repellent is strongly recommended.

The visitor center and rest rooms are accessible to the disabled, but only about 250 yards of trail near the center are paved. Some roadside birding is possible from the miles of road and seven little-used parking lots in the park.

10. Kaneohe Marine Corps Base Hawaii

FEATURES:
• Spectacular booby colony
• Seabirds

FACILITIES: None

One of the most unlikely birding spots anywhere is the Kaneohe Marine Corps Base Hawaii on the windward shore of O'ahu. Situated improbably between the marines' rifle range and the sea is a large nesting colony of Red-footed Boobies. The thirteen hundred birds show almost no fear of humans. They don't even seem to get nervous, unless you approach within three or four feet of a nest. The air is filled with boobies on the wing, the calls of boobies, and the heady aroma of lots of big birds living in a warm, wet environment. A visit to this colony is totally overwhelming.

But that's not all. Less than a mile offshore from the booby colony is Moku Manu Island, a seabird sanctuary that is home to many nesting seabirds, including 10,000 to 20,000 breeding pairs of Sooty Terns, 5,000 pairs of Wedge-tailed Shearwaters, 1,000 to 3,000 pairs of Brown Noddies, 300 to 600 pairs of Red-footed Boobies, fewer than 100 pairs each of Black Noddies, Brown Boobies, and Gray-backed Terns, and fewer than a dozen pairs of Bulwer's Petrels and Masked Boobies. The table below

Table 5
Seasonal Occurrence of Breeding Birds at Moku Manu Island

BIRD	J	F	M	A	M	J	J	A	S	O	N	D
Wedge-tailed Shearwater					———	———	———	———	———	—		
Bulwer's Petrel						———	———	———				
Sooty Tern	———	———	———	———	———	———	———	———	———	———	———	—
Brown Noddy	———	———	———	———	———	———	———	———	———	———	———	—
Black Noddy	———	———	———	———	———	———	———	———	———	———	———	—
Christmas Shearwater			———	———	———	———	———	—				
Red-footed Booby	———	———	———	———	———	———	———	———	———	———	———	—
Brown Booby	———	———	———	———	———	———	———	———	———	———	———	—
Masked Booby	———	———	———	———	———	———	———	———	———	———	———	—
Gray-backed Tern			———	———	———	———	———	———	———	———	—	

Compiled from observations, interviews, and information in Richardson and Fisher, "Birds of Islands off Oahu," *The Auk,* Vol. 67, No. 3, July 1950, *Hawai‘i Wildlife Plan,* State of Hawai‘i, 1983, and *Seabirds of Hawaii* by Harrison, 1990.

shows when these birds are present on Moku Manu. A single pair of Great Frigatebirds has nested on the island, and White-tailed Tropicbirds and Laysan Albatrosses often visit the area. And to make the birder's fantasy complete, there are ponds on the air station where Black-necked Stilts nest.

The juxtaposition of the Kaneohe Marine Corps Base Hawaii and a rare colony of boobies may not really be much of a surprise. These birds nest in bushes or low trees where they are vulnerable to human interference and predation by rats, cats, and dogs. They can nest successfully only where they are actively protected, as they are at the air station. This protection results from the severe restrictions on access that the Marine Corps imposes.

Perhaps now the picture is beginning to focus. This is a great birding spot, but it is very difficult to arrange a visit, especially for visitors who are in Hawai‘i only for a short time. There are two ways you might be able to see the booby colony. First, write to the Hawai‘i Audubon Society for an annual schedule of their field trips. The schedule is usually published in January or February each year (see the appendix for more information). Audubon usually visits the colony at least once a year. These visits book up fast, so make your reservations as soon as possible. Some-

An Audubon field trip to Kāne'ohe

times the date of the visit must be changed on short notice to accommo-
date Marine Corps operations, so be prepared for disappointment.

You can also write to the Marine Corps and request an individual
tour. Requests are processed on a case-by-case basis, and there is a limited capability to support such tours. You may have a better chance of
success if you are available on several days and you let the marines pick
the exact date and time. Once again, changes may be made on short
notice. Address your request to: Director, Joint Public Affairs, United
States Marine Corps, Kaneohe Marine Corps Base Hawaii, Kāne'ohe Bay,
Hawai'i, 96863–5001.

If you are lucky enough to arrange a visit, you can reach the Marine base from Honolulu by taking the Pali Highway (Route 61) all the way to its end in the town of Kailua. At the T intersection, turn left on Kalāheo Avenue and follow it about 2.5 miles to the intersection with the H-3 Freeway. Take the freeway a short distance to the right to the air station entrance.

If you visit the booby colony, be sure to take a camera. You will be able to approach the nesting boobies so closely that no telephoto lens will be required to get good photos.

Additional birds that might be seen on the sprawling military installation include Mallard, Hawaiian Coot, Common Moorhen, Cattle Egret, and Black-crowned Night-Heron near the ponds, and Common Myna, Zebra Dove, Spotted Dove, Red-crested Cardinal, Northern Cardinal, Red-vented Bulbul, and Kōlea around developed areas. Black Noddies can often be seen feeding in nearby Kāne'ohe Bay.

11. Kualoa Regional Park

FEATURES:

- Common Waxbills
- Wetland birds
- View of scenic islet

FACILITIES:

One of the most scenic developed parks anywhere has got to be Kualoa Regional Park along the windward coast of O'ahu. The standard park amenities are all available: lots of lawn, rest rooms, drinking fountains—even showers for the swimmers and campers. But the real attractions of this spot are the vistas. Look *mauka* and you see the steep windward flank of the Ko'olau Range towering nineteen hundred feet over the flat Kualoa Peninsula, the peaks often obscured by clouds that keep this area moist and green. Turn around and you face one of the prettiest offshore islets in the state, Mokoli'i or Chinaman's Hat.

Just when you think nothing could improve this perfect setting, a Cattle Egret glides in to hunt for insects on the lawn, joining the Kōlea. As the egret alights, it startles a flock of Common Waxbills, finches so tiny that they are hidden by the mowed blades of grass as they forage for seeds. The waxbills depart for a nearby tree already occupied by a Northern Cardinal and a Red-crested Cardinal who relinquish their perches. The Red-crested Cardinal, accustomed to the begging routine of the park, heads for the greener pastures of the parking lot to join House Sparrows, Spotted Doves, Zebra Doves, Common Mynas, and Red-vented Bulbuls.

Meanwhile, the Northern Cardinal joins its mate in the tangle of trees surrounding the pond at the far end of the park. These trees are noisy with the calls of the cardinals as well as Japanese White-eyes and White-rumped Shamas. Out in the pond, a Black-crowned Night-Heron seems to ignore human park users who peer through the vegetation, while a Black-necked Stilt calls in alarm at the first approach of any two-legged creatures without wings. Mongooses are also alarmed at the approach of humans, and slink toward cover. Flocks of Nutmeg Mannikins move a safe distance down the nearby ocean shoreline to another tall clump of grass. Out over the ocean, Black Noddies, Wedge-tailed Shearwaters, and White-tailed Tropicbirds scan the waves for their next meal. Offshore, a thousand pairs of Wedge-tailed Shearwaters share Mokoli'i Islet with small numbers of White-tailed Tropicbirds and Red-tailed Tropicbirds.

If this sounds like a birder's delight, it is likely to get even better in the future. Right now, all the land birds described above are near certainties at Kualoa, but the wetland birds are sometimes absent. Recently the County and the State have reached agreement on more intensive management of the pond area, with initial activities to include predator control and removal of trash and undesirable vegetation.

Most of the birds in the park can be viewed from the parking lot or the lawn areas. To look for wetland birds, follow the long parking area all the way to the end, cross the grass to the low stone wall, and climb up on it for a view over the vegetation between the wall and the pond.

The park is at the 30.6 mile point on Highway 83. Except for the pond area, all parts of the park including the rest rooms have good access for the disabled. Park hours are from 7:00 A.M. to 8:00 P.M.

12. Kahuku Area Wetlands

FEATURES:
• Wetland Birds

FACILITIES:

Near the northernmost point of O'ahu are wetland areas that provide essential habitat for wetland birds—and excellent viewing for birders. These areas include the James Campbell National Wildlife Refuge and some adjacent aquaculture ponds. Both are near the fifteen-mile marker on the Kamehameha Highway (Highway 83), just northwest of the town of Kahuku.

The James Campbell National Wildlife Refuge was established in 1977 on 142 acres of privately owned land in two parcels leased from the James Campbell Estate. One parcel contains the spring-fed Punamanō Pond. The other parcel, known as the Ki'i Unit, contains settling basins used by the sugar mill at Kahuku before it shut down in 1971. The refuge is being restored and enhanced to provide good nesting and feeding habitat for wetland birds.

Access to the refuge is restricted, and there are currently no plans to create any public viewing areas. However, you can arrange a visit by calling the refuge office to obtain a special use permit. Your permit should be requested at least two weeks before you plan to visit, and preferably one to two months in advance (see appendix). The Hawai'i Audubon Society also organizes visits to the refuge occasionally. The refuge is closed from February 1 to July 30 during breeding and nesting season.

Adjacent to the national wildlife refuge are aquaculture ponds used to grow shrimp and tilapia fish. The ponds are privately owned, but there is excellent viewing from the wide shoulder of the highway. There is plenty of room to park and walk along the fence, or view birds from your car. The birds are sometimes a long way from the road, and thus a spotting scope could come in handy. The James Campbell Estate owns these ponds, and leases them to aquaculturists. In 1995 the ponds were drained and not in use, and a new operator was being sought. The future

Black-necked Stilt

of birding at these ponds will obviously depend on their continued operation.

Hawai'i is home to four endangered wetland birds, and all four of them may be found in the Kahuku area. The Koloa duck and Hawaiian Coot are endangered species that are endemic to the Islands. Subspecies of two other wetland birds are unique to the Islands and are also endangered: the Black-necked Stilt and Common Moorhen. Another bird native to Hawai'i that may be seen at Kahuku is the Black-crowned Night-Heron, called 'Auku'u in Hawaiian.

Sometimes biologists don't know how a bird has gotten to Hawai'i, and they aren't sure whether to classify it as an introduced bird or a new "native" that has arrived on its own. That's the case with the Fulvous Whistling-Duck. This species first appeared in the area in 1982 and has been breeding. Was it introduced illegally, or did it wander to Hawai'i on its own? Nobody knows for certain. Watch for these birds at Kahuku. Their large size and tall upright stance distinguish them from other ducks, even at a distance.

Wetlands have become scarce in Hawai'i, so migrant waterfowl and wetland birds tend to concentrate in the few areas—like Kahuku—that remain. During the winter look for Kōlea, Ruddy Turnstone, Sanderling,

Wandering Tattler, Northern Pintail, and Northern Shoveler. Less common are Pectoral Sandpiper, Sharp-tailed Sandpiper, and Long-billed Dowitcher. This is a good place to see other uncommon migrants and stragglers, including ducks and wading birds. Watch for the Great Frigatebird overhead and Pueo soaring over grassy areas. There are Mallards here that breed with the Koloa, but they are probably introduced. Other introduced birds include Cattle Egret, Ring-necked Pheasant, Spotted Dove, Red-vented Bulbul, Common Myna, Common Waxbill, Nutmeg Mannikin, and Red Avadavat.

13. Ka'ena Point

FEATURES:
- Marine birds
- Rare dune ecosystem
- Native plants

FACILITIES:

Ka'ena Point is a relatively remote State Natural Area Reserve where you can view marine birds and rare Hawaiian plants. This area is one of the state's best examples of coastal lowland and dune ecosystems. Formerly popular with off-road vehicle enthusiasts, it was closed to ORV use in 1988 to help the rare plants in the area recover. The recovery has been quite successful, and today many species of native dune plants are making strong comebacks. The State of Hawai'i publishes a good brochure describing and illustrating several native plants that can be seen in the area. The brochure can be obtained from the Natural Area Reserves System (see appendix).

The area is best known for its endangered plants, but it is a good place to see marine birds as well. Ka'ena Point juts out into the ocean, and going there is almost like being at sea. From spring to fall, Wedge-tailed Shearwaters and Brown Noddies may be seen. Red-footed Boobies are present on O'ahu year-round and might be seen at any time of year. Brown Boobies might be seen during the summer. Look for Great Frigatebirds soaring overhead year-round. Kōlea can be seen from August to

April. During winter the Laysan Albatross is a fairly common visitor. In fact, a few have attempted to nest at Ka'ena Point in recent years, but most of the young fall prey to cats, mongooses, and vandals. Introduced birds to watch for include Common Myna, Red-vented Bulbul, and Spotted Dove.

Lucky visitors may also see marine mammals around Ka'ena Point. Humpback whales are often seen off the point during their wintering season in these waters. Dolphins can sometimes be seen in the early morning at Yokohama Bay by the mouth of Kaluakauila Stream in adjacent Mākua–Ka'ena State Park. On rare occasions, solitary Hawaiian monk seals are seen basking on the rocks at the point. Be careful not to disturb the endangered monk seals; stay at least a hundred feet away.

This westernmost point of O'ahu can be reached from its northern (Mokulē'ia Beach) side or from the southern (Wai'anae) side. Roads leading to both sides of the area are named the Farrington Highway (Highway 930) although they do not connect at Ka'ena Point. You can hike in about 2.5 miles to the point from either side. The northern side is heavily used by off-road vehicle enthusiasts between the trailhead and the reserve boundary, especially on weekends. The southern route starts in a state park, is a little more scenic and much quieter, and may be a better route for those seeking quiet contemplative activities. If you approach from the north side in winter, check the Dillingham Airfield, located just before the end of the road. Laysan Albatrosses have explored this area for a nesting site in recent years.

If you approach from the south side, be sure to stop at Kea'au Beach Park, 4.3 miles before the gate to Ka'ena Point. The lawn at Kea'au is a good place to look for Red Avadavats.

At either Ka'ena trailhead you will be leaving your car in an isolated part of rural O'ahu where automobile break-ins are very common.

There are drinking fountains at Kea'au Beach Park. Telephones and rest rooms are available at Mākua–Ka'ena State Park. There is no water available along the trail and there is no shade, so take plenty of water and sunscreen, and plan for a hot hike. You will certainly be rewarded with solitude. Even on weekends only a handful of anglers and hikers visit the point.

14. Rainy Day Birding—Bishop Museum

FEATURES:

• Historical perspective on Hawaiian birds

FACILITIES:

If rain keeps you from visiting the forest trails around Honolulu, you can go birding at the Bishop Museum—and see extinct birds at that! This major museum has exhibits on many aspects of natural and cultural Hawai'i and Polynesia. Of particular interest to the birder is the world's largest collection of authentic Hawaiian feather work, the grand cloaks and helmets fashioned for Hawaiian leaders from millions of bird feathers. Prominent in this feather work are the feathers of the 'I'iwi and 'Ō'ō species. There are also a few mounted birds that have become extinct since these specimens were collected—grim reminders of the biological riches that we have lost on these Islands.

Allow plenty of time to visit the Bishop Museum. There is a lot to see besides birds and feather work, including a planetarium and a special hands-on section for children. You could easily spend half a day at the museum, and you might want to spend an entire day, lunching at the museum café or picnicking on the grounds.

The feather work and extinct Hawaiian birds are located on the first floor of Hawaiian Hall. The second and third floors of this very old building are not wheelchair-accessible, but all other areas of the museum have good access.

The Bishop Museum is located at 1525 Bernice Street in Honolulu, very near the intersection of the H-1 Freeway and Highway 63 (Likelike Highway). The museum is open from 9:00 A.M. to 5:00 P.M. daily. An admission fee is charged.

Other O'ahu Birding Spots

Several spots around O'ahu, described below, offer the possibility to combine birding with other recreation.

Sand Island State Park

This is a 140-acre landscaped beach park right in Honolulu. Seabirds are almost always visible from this park, but they are often too distant for easy identification. In winter Pomarine Jaegers may be seen. Shorebirds and urban birds can also be seen. To reach the park, take the Nimitz Highway (Route 92) to Sand Island Road (Route 64). Follow Sand Island Road all the way to its end.

Hanauma Bay

This marine life conservation district, just east of Honolulu, is famous for its snorkeling. In recent years it has also become famous for millions of tourists who are literally loving the area to death. If you visit this underwater park and marine life conservation district, notice the birds that take advantage of the ample handouts. At the lookout area near the parking lots House Sparrows, Red-vented Bulbuls, and Red-crested Cardinals are frequent beggars. They are so tame that you can often get good photos with a standard lens.

Down at the beach, anyone passing out bread crumbs is soon surrounded by Rock Doves and House Sparrows. The Rock Doves nest in niches in the rocky cliffs toward both ends of the beach. Seeing these birds in a natural setting gives a better appreciation of the term Rock Dove. They really seem at home.

Green sea turtles often visit this bay. Sea turtles are notorious for eating plastic bags—which can be fatal to them—perhaps because the bags look like edible jelly fish. Unfortunately, many visitors carry fish food into the water in plastic bags, become mesmerized by the fish, and lose their bags. So if you see a plastic bag in the water while you are snorkeling, help a turtle by retrieving the bag and disposing of it properly.

Sacred Falls State Park

Along the windward coast of O'ahu is a scenic spot where an easy-to-moderate hike is rewarded with many introduced birds and a splendid waterfall. The hike through Sacred Falls State Park takes you past old cane fields and into a shady canyon with walls sixteen hundred feet high. Sacred Falls, or Kaliuwa'a Falls as they are more properly called, plunge eighty-seven feet into a pool surrounded by sheer cliffs and green forest.

The well-marked park is one mile south of Hau'ula town on Highway 83, or about four miles south of the Polynesian Cultural Center. The distance from trailhead to falls is 2.2 miles. For the first 1.2 miles the trail follows an old cane road, which can be hot during the afternoon because there is little shade. Along this part of the trail you can expect to see Red-vented Bulbul, Spotted Dove, Zebra Dove, Common Myna, House Finch, House Sparrow, Nutmeg Mannikin, Northern Cardinal, and Red-crested Cardinal. From August to April you will likely see Kōlea in the grassy area at the trailhead. One bird that you will hear but probably will not see is the Japanese Bush-Warbler. Its song has been described as flutelike, with a long whistle followed by several quick notes. The bird's Japanese name, Uguisu, sounds something like its song (ooo-ooo-ooo-goo-EEE-soo). The singing is seasonal, and stops almost entirely between September and December. Two uncommon birds to watch for are Common Waxbill and Orange-cheeked Waxbill.

The last part of the trail follows a narrow canyon to the falls. It is cool and shady along this part of the trail, with a few different birds to watch for. The Japanese White-eye might be seen anywhere along the trail, but seems more common in the canyon. The White-rumped Shama is common in the canyon as well. Listen for the singing Shamas, competing with the babbling of Kaluanui Stream: very nice sounds indeed. Finally, look for White-tailed Tropicbirds soaring around in the canyon very near the falls.

The trail is popular among visitors and locals, so the best birding is early in the morning when there are few other hikers. The trail is not accessible to the disabled.

5
Kaua'i

KAUA'I IS A GOOD ISLAND for birders. Two of the most irresistible birding spots in the state are on this island, and they are such scenic places that you'd want to visit them even if you weren't interested in birds. These are Kīlauea Point on the island's north shore and Kōke'e State Park in the mountainous interior. The rest of the island is scenic, too. It is called the Garden Isle because of the lush tropical vegetation that abounds on much of the island. This greenery is the result of abundant rain on the windward side and in the mountains, coupled with the deep topsoil that has had time to form on this oldest of the main islands.

The rain and the vegetation have worked together to protect some of Hawaii's last endemic forest birds. The Alaka'i Swamp in the mountainous interior is such an impenetrable, muddy tangle that it has been disturbed less than most bird habitats in the state. This area is also high enough—at about four thousand feet—that there are very few mosquitoes. The result is that some of the rarest birds in the state—or anywhere on the planet, for that matter—make their last stronghold in and around the Alaka'i. Although you should not expect to glimpse the rarest of these birds, your visit to Kōke'e Park on the edge of the Alaka'i will be rewarded with sightings of many endemic forest birds.

Hurricane 'Iniki

On September 11, 1992, Hurricane 'Iniki hit Kaua'i with sustained winds of over 130 miles per hour, and gusts of 175. The human suffering was well documented by the media: most island residents suffered damage to their homes, and many were left homeless. Services were interrupted, in some cases for months. As tragic as the event was for the people of Kaua'i, the impacts were perhaps even more severe for the island's natural systems. Large areas of forest, particularly at middle elevations, were

stripped of foliage. Trees were snapped off or uprooted. In some areas
the winds tore away all vegetation, leaving bare dirt.

We might expect the plants and animals of Hawai'i to be adapted to
hurricanes and to have mechanisms for surviving such events. After all,
Hurricane 'Iwa hit Kaua'i just ten years earlier, and presumably hurricanes have swept over the islands regularly during the millions of years
these points of land have existed. In fact, there are indications that native
plants are adapted to hurricanes and the conditions they leave behind.
Koa seedlings seem to sprout vigorously after hurricanes. The young
plants apparently favor the sunny areas created when winds topple
mature koa trees and open up the forest canopy. But today there are
many alien plants ready to compete with native species. As long as the

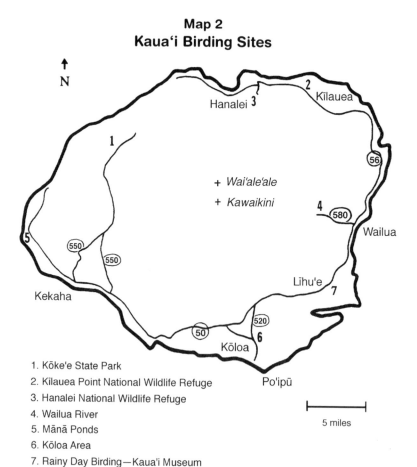

Map 2
Kaua'i Birding Sites

1. Kōke'e State Park
2. Kīlauea Point National Wildlife Refuge
3. Hanalei National Wildlife Refuge
4. Wailua River
5. Mānā Ponds
6. Kōloa Area
7. Rainy Day Birding—Kaua'i Museum

forest is intact these aliens may penetrate slowly, but hurricanes spread alien seeds and open up the canopy, giving aliens as well as natives a place to germinate. The aliens may outcompete natives and expand their range into the native forest. As the forest recovers after Hurricane 'Iniki, it may be transformed, with more alien plants growing in more places than before.

Birds face special perils during and after hurricanes. Seabirds may fly out to open ocean to escape from the path of the storm, but forest birds don't have that option. Many were killed by Hurricane 'Iniki, and others starved in the stripped forests that remained after the storm. Some honeycreepers were observed venturing down to lower elevations in search of food after the hurricane. For birds such as the 'I'iwi, with no natural immunity to avian malaria, one trip down into the mosquito zone can mean death.

Before human occupation of the Islands and the resulting habitat modification, suitable habitat for forest birds was probably much more extensive. Forest birds might have found some areas that were spared from a hurricane's effects where they could survive until other forest areas recovered. Bird populations were certainly larger before humans arrived, and that meant more birds would survive a hurricane and be available to help the population recover.

There is no going back to a time when native species had less competition as they recovered from the effects of hurricanes. The reduced ability of Hawaiian plants and animals to rebound in the face of habitat modification and introduced species is a reflection of the changes we have brought to the Islands. Every time a hurricane hits, we are bound to be left with a little less of what existed before.

Visiting Kaua'i After the Hurricane

Despite the effects of Hurricane 'Iniki, Kaua'i is still one of the better islands for visiting birders. Years after the storm there are still reminders of its destruction, such as toppled trees and shuttered condominiums, but these are exceptions. Most of the island is as lush and green and hospitable as it was before.

Visitor accommodations on Kaua'i are more evenly distributed than on any of the other islands. Sunny Po'ipū on the south shore offers many expensive condominiums set in lush grounds, and several hotels. They were hit hard by 'Iniki, but most have reopened. A few towns on the east shore offer hotel rooms and condominiums in a range of prices.

Princeville on the north shore is most popular with golfers and rain lovers. Homes are often for rent in picturesque Hanalei on the north shore and out in Kekaha on the south.

There is no clear first choice of lodging for birders, but the Po'ipū area may have a slight edge over other parts of the island. Although accommodations at Po'ipū tend to be more expensive than along the east shore, the climate is much drier and sunnier. Since you may get very wet at some of the birding spots on the island, a dry base of operations will be appreciated. It is usually possible to attract a variety of birds to your condominium lanai at Po'ipū. The area also offers fine beaches with moderate snorkeling.

There is another good lodging choice on the island. I heartily recommend that visitors spend at least a night or two at Kōke'e State Park. The conifers and the cool mountain air may not seem much like Hawai'i, but the birding, hiking, scenery, and relative solitude are unparalleled.

1. Kōke'e State Park

FEATURES:
- Endemic forest birds
- Scenic views

FACILITIES:

Every Hawaiian island offers special places for birders, and Kōke'e is perhaps the most special among them. Although hurricanes and mosquitoes have taken their toll, you are likely to see as many endemic forest bird species at Kōke'e as any other place in the state. The area also offers spectacular views of Waimea Canyon, the Kalalau Valley, the Alaka'i Swamp, and the usually mist-shrouded Wai'ale'ale, reputed to be the rainiest spot on earth. Finally, the park offers the amenities to make a birding trip enjoyable: a fine little natural history museum, many good hiking trails, a restaurant, and housekeeping cabins.

The forests of Kōke'e and the Alaka'i sustained less hurricane damage than some of the lower-elevation forests in 1992, although downed trees are common sights in the area. The forest bird species you should

Housekeeping cabin at Kōke'e

be able to see at Kōke'e include (in approximate order of abundance) 'Apapane, Kaua'i 'Amakihi, 'Elepaio, 'I'iwi, and 'Anianiau. Seeing an 'I'iwi at Kōke'e used to be nearly a sure thing, but that was before Hurricane 'Iniki. Now this species is uncommon in the area. With a great deal of luck, you might see an 'Akeke'e (formerly Kaua'i 'Ākepa), or an 'Akikiki (formerly Kaua'i Creeper).

Endangered Birds and the Alaka'i

Adjacent to Kōke'e State Park is the Alaka'i Wilderness Preserve, whose principal feature is the Alaka'i Swamp. This area is famous for being the last stronghold of several critically endangered forest birds, including the Kaua'i 'Ō'ō or 'Ō'ō 'Ā'ā, 'Akialoa, Nukupu'u, 'Ō'ū, Kāma'o, and Puaiohi. Most birders would love to get a glimpse of these rare creatures, but there are compelling reasons not to try. First, and sadly, several of them are probably already extinct. The 'Akialoa has not been sighted since the 1960s. Only a single 'Ō'ō has been seen since Hurricane 'Iwa devastated the island in 1982, and even this lonely bird has not been detected since 1988, despite determined efforts. Before Hurricane 'Iniki the others were clinging tenuously to existence, with only the Puaiohi surviving in numbers that offered any real hope of avoiding extinction. A bird survey of the Alaka'i in 1994 turned up seventeen Puaiohi, but none of the other six endangered species. In 1995 a Nukupu'u was reported from the Ala-ka'i Swamp Trail. Lack of sightings doesn't necessarily mean the 'Ō'ū and Kāma'o are extinct, but it suggests extremely low numbers.

If any of these birds are to survive, an undisturbed habitat is prob-ably essential. Whenever people venture into the Alaka'i Swamp, there is the possibility that alien weed seeds will inadvertently be carried along on the soles of boots. So little is known about the biology of these birds that we don't know for certain whether mere human presence might cause nest abandonment or other disasters.

In addition to survival of the birds, there is survival of the birder to consider. The remotest part of the Alaka'i Swamp is an inaccessible and unforgiving environment. The only way to reach the heart of the Alaka'i is to slog through mud at least thigh deep. The few trails that penetrate this area are difficult to follow, and the usual foul weather might delay or prevent any attempts to rescue lost birders. Camping is not permitted in the swamp, so any birding foray must be a one-day trip that leaves little time for birding once you get into the area.

A few of the trails described in this section will take you into the edges of the swamp. They probably give you as good a chance as any of spotting endangered forest birds. For the survival of any remaining birds, and for your own comfort and safety, I do not recommend venturing farther.

Planning Ahead

If you plan to visit this area, there are a few things you will want to do in advance. First, try to arrange spending a night or two in the park. The best birding and the best chances for cloudless views are early in the morning, and Kōke'e is a long drive from any other lodging. Your best bet is one of the reasonably priced housekeeping cabins. These cabins, operated by a concessionaire in the park, are rustic and some of them are quite old, but they are generally clean and quite adequate for "rough-ing it" in the mountains. They are equipped with full kitchens, linens, and wood stoves for warmth. Most basic kitchen utensils are provided, but the sharp blade and can opener on my trusty Swiss army knife have come in handy more than once. The cabins are very popular and it is essential to reserve one as far in advance as possible, preferably four to six months. Holiday periods can book up a year in advance. The maxi-mum stay is five nights. The cabins are elevated, so they do not have good access for the disabled. To reserve a cabin, contact Kōke'e Lodge, P.O. Box 819, Waimea, Kaua'i, HI 96796, (808) 335-6061.

There is also a camping area in the park. A permit from the State is required. Keep in mind that the nights can be quite cool, especially in the winter, and it can be very rainy. To reserve a campsite, contact the Division of State Parks in Līhu'e. There is also a Division of State Parks office in Honolulu where permits can be obtained. See the appendix for addresses and phones.

The lodge at Kōke'e contains a restaurant, a small store, and rest rooms. If you plan to spend one or more nights at Kōke'e, you will want to bring your own groceries; the selection in the store is quite limited. Lodge hours are 9:00 A.M. to 3:30 P.M. but are subject to change. The building has good access for the disabled.

Getting There

To reach the Kōke'e area, take Highway 50 from Līhu'e toward Waimea. From Highway 50 you will turn *mauka* on Highway 550. There are two roughly parallel branches of Highway 550 that proceed from Highway 50 toward the park. The first branch intersects Highway 50 at the latter's 23.1-mile point in the town of Waimea. This is Waimea Canyon Drive. The highway sign will direct you to the second branch, which intersects Highway 50 in the town of Kekaha at the 26.3-mile point. Most people take the first branch, Waimea Canyon Drive. It is more direct and thus the road is steeper, but it is in somewhat better condition than the other route. There are several turnouts with views of the lower Waimea Canyon. For birders, this route offers the possibility of seeing a White-tailed Tropicbird over the canyon, but there are other canyon overlooks that provide the same opportunity. You may also see Pueo over the cane fields.

The second branch of Highway 550 is longer and is less frequently used, so there is less traffic to frighten birds. It is not as well maintained, and in places grass hangs over onto the pavement, providing cover for Erckel's Francolins and food for Nutmeg Mannikins and Chestnut Mannikins. Zebra Doves and Spotted Doves often stand around on the roadway, and Northern Cardinals and Mockingbirds inhabit the dry gulch bordering the road. There are definitely more birds to be seen on this route. You may want to take one route on your way to Kōke'e and the other coming back down so that you don't miss anything.

Individual birding spots in the Kōke'e area are described below. If you hike any of the trails, be sure to inquire about current trail conditions at the museum, get a trail map, and register at trailheads.

Kōke'e Natural History Museum

This tiny nonprofit museum near the park headquarters is a good place to start your visit to the area. The displays give good background on the birds, plants, topography, and weather of the area. A very informative display on hurricanes is a recent addition. Free trail maps are available, better ones are for sale, and a good selection of natural history books is available. The museum is open from 10:00 A.M. to 4:00 P.M. daily, and a small donation is requested. There is good access for the disabled.

Kalalau Lookout

The Kalalau lookout is probably the most-visited spot in the park because of the breathtaking clear-weather view down into the Kalalau Valley. The lookout is also highly recommended to birders: despite the thousands of people who visit this spot every week, it is one of the best birding areas in the park. For birders eager to see endemic species, it is one of the best easily accessible birding spots in all of Hawai'i.

Try to visit very early in the morning, when you have the place nearly to yourself. Early morning is also the time when you are most likely to get a clear, cloudless view into the Kalalau Valley. As you drive to the lookout, watch for Erckel's Francolin and Ring-necked Pheasant along the roadsides. Around the parking lot, you may see Common Myna, Kōlea, House Finch, and Nutmeg Mannikin.

Nearby are the rest rooms, surrounded by a thicket of introduced understory plants and tall 'ōhi'a. In this dense vegetation look for 'Elepaio, 'Anianiau, Kaua'i 'Amakihi, Hwamei, and White-rumped Shama.

A short paved trail takes you to a level grassy area with spectacular clear-weather views down into the amphitheatre of the Kalalau Valley. If the view is obscured by clouds, watch the birds in nearby trees for a while. Sometimes the clouds break briefly to give views into the valley. White-tailed Tropicbirds can usually be seen soaring around in the valley below. If the weather is clear these birds will be surprisingly easy to pick out in the immensity of the valley, their white plumage contrasting with the reds and greens of the landscape. Also watch for feral goats in the valley below, particularly on ridgetops where some grass grows. Finally, shorten your gaze to the treetops just below you on the steep hillside. Watch the tops of these trees for 'Apapane, Kaua'i 'Amakihi, and perhaps an 'I'iwi, 'Anianiau, or even an 'Akeke'e.

The grassy area at the overlook is dotted with 'ōhi'a trees. In years of visits, I have never failed to see 'Apapane working through these trees, sipping nectar from the lehua blossoms. If you sit quietly beneath the trees you may be able to observe an 'Apapane at very close range, both of you oblivious to any visitors at the other end of the overlook area, peering down into the valley.

Pu'u o Kila Lookout

At the end of the paved highway is the Pu'u o Kila Lookout, perched on the knife edge of a ridge separating the Alaka'i Swamp from the Kalalau Valley. This overlook presents a slightly different view down into the valley, and affords views the other direction into the Alaka'i Swamp. This is also the trailhead for the Pihea Trail (see below).

This can be an excellent spot for birding if the weather cooperates. The best place to look is in the treetops below the overlook on the inland, Alaka'i side of the ridge. On this side there is protection from the wind, and birds are not viewed against a glary sky. 'Apapane and Kaua'i 'Amakihi may be seen. 'Anianiau, 'I'iwi, and 'Akeke'e are slim possibilities. Below the canopy you are likely to see 'Elepaio, and with extra-ordinary luck you might see an 'Akikiki. The overlook itself is often buffeted by strong winds. At these times, hapless birds often blow by at high speed, flying sideways against the wind, sometimes silhouetted against a gray, cloudy sky. These are certainly not the best birding conditions.

Around the parking lot, species to look for include House Finch, Common Myna, Kōlea, and Nutmeg Mannikin. In the understory look for White-rumped Shama, and watch for Hwamei skulking near the ground.

Pihea Trail

The Pihea Trail offers one of the more strenuous hikes in the park, as well as some of the best opportunities to see 'Akeke'e and 'Akikiki. The trailhead is at the Pu'u o Kila Lookout. The trail follows the ridgetop for about one mile, along a scar created in the 1950s by an ill-conceived and eventually abandoned road project. There may be Kōlea along the trail in open areas, and you are likely to see 'Elepaio as you pass through wooded sections of the trail. Also watch for Hwamei and White-rumped Shama. Where you have views into the treetops below you on the Alaka'i side, you may spot any of the more common endemic forest birds.

After about three-quarters of a mile, the trail begins to get muddy. Also, there are steep places where you will need to pull yourself up by grasping trailside branches and exposed roots. At about the one-mile point the trail turns and descends toward the Alaka'i Swamp, with more

The Pihea Trail

steep, slippery stretches. Where the terrain levels out, boardwalks have been built across bog areas. Before the boardwalks were built in the early 1990s, hikers would skirt around the deep mud, disturbing the trailside and making the trail wider and wider at these points. Some of this mud was waist deep or more. If you never slogged through the old Pihea Trail, you can't fully appreciate the new boardwalk.

The trail continues for another mile to an intersection with the Alaka'i Swamp Trail, then continues for more than a mile to its end at the Kawaikoi Stream Trail. All the boggy stretches along this once-grueling trail have now been replaced with boardwalks.

You may want to hike the Pihea Trail until it starts to descend into the Alaka'i, and then sit quietly along the trail at a place where you have good views of treetops and understory. Wait for feeding groups of birds to come to you, for curious 'Elepaio to approach, for skinks and rats and other forest creatures to resume their normal activities around you. This is probably your best bet to see an 'Akikiki.

Kaluapuhi Trail

This is a fairly good birding trail, with the added benefits of being mostly level and mostly free of mud. The trail starts at the 17.2-mile point on the *mauka* or inland side of Highway 550, passes for two miles through the forest, and comes back out on the highway past the Kalalau Lookout. For the first mile the trail is usually well maintained. At that point you will come to a fork, and both branches quickly reach stretches that can be hard to follow. The right branch, not a formal trail, dead-ends at a brushy overlook. The left branch returns to the highway, but can be quite over-grown and rather soggy in wet weather.

Along the trail look for 'Apapane, 'Elepaio, Kaua'i 'Amakihi, 'Ani-aniau, Japanese White-eye, White-rumped Shama, and Hwamei. The fork in the trail is in a thicket of strawberry guava, and there are many intro-duced berry vines nearby. These provide food for the House Finch.

Awa'awapuhi Trail

This hike is highly recommended for people in good physical condition. The area offers very good birding, and if you make it the full 3.1 miles to the end of the trail you will be rewarded with one of the most stunning vistas you will ever see. The catch is that you have to work for all this, and work hard. The trail descends from the highway to a lookout that is eleven hundred feet lower in elevation. The walk down is great, but the long, gradual climb back up the ridge is rather strenuous. Do not attempt

Map 3
Kōke'e Area Trails

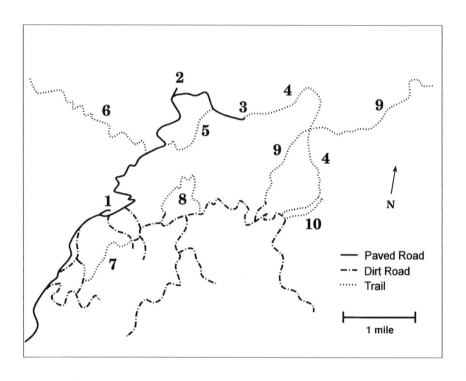

1. Kōke'e Natural History Museum
2. Kalalau Lookout
3. Pu'u o Kila Lookout
4. Pihea Trail
5. Kaluapuhi Trail
6. Awa'awapuhi Trail
7. Halemanu-Kōke'e Trail
8. Pu'u ka 'Ōhelo/Berry Flat Trail
9. Alaka'i Swamp Trail
10. Kawaikōī Stream Trail

Note: Not all Kōke'e trails are shown on this map.

this hike unless you are in good physical condition. Carry food and water.

The trailhead is just past the seventeen-mile marker on the *makai* or ocean side of Highway 550 at a large parking area. The trail descends from 'ōhi'a forest through 'ōhi'a-koa forest and into a more arid forest scrub. Most of the mid-elevation forests of this type have been destroyed or greatly disturbed, so this trail offers a rare opportunity to see many native and endemic plants. At one time, many plants were labeled with numbered posts and described in an excellent trail guide available from the State office in Līhu'e or at the Kōke'e museum. Unfortunately, vandals and Hurricane 'Iniki have destroyed nearly all the posts. If they are replaced, the Awa'awapuhi Botanical Trail Guide will once again be a very useful and informative publication. Ask at the museum for the current status.

You may be surprised to see endemic birds in this forest all the way down to the end of the trail at an elevation of 2,980 feet, which is well into the mosquito zone. I have seen 'Apapane, 'Elepaio, and even 'Akeke'e within half a mile of the end of the trail. Either these birds are flirting with death from mosquito-borne avian diseases, or they have developed some immunity to introduced diseases.

At the lower elevations you may also see Northern Mockingbird, Northern Cardinal, and Nutmeg Mannikin. You will probably hear the cackling call of Erckel's Francolins carrying across the valleys. At the end of the trail you are very likely to see feral goats. I have seen as many as nine at a time in this area. Watch for White-tailed Tropicbirds soaring across the valley below you.

Someday birders may also see a breeding population of Nēnē in this area. In 1995 State wildlife officials, encouraged by the success of Nēnē flocks at Kīlauea and near Līhu'e, released about thirty birds in Nu'alolo Valley.

Trailheads Off the Highway

Several trailheads are off the paved highway on dirt side roads, including trailheads for Halemanu–Kōke'e, Pu'u ka 'Ōhelo/Berry Flat, Alaka'i, and Kawaikōī Trails. You can usually drive to the nearer trailheads in a passenger car, but do not attempt to drive off the pavement during very wet weather unless you have 4WD. You can ask about road conditions at the museum. To reach these trailheads, go 0.1 miles past park headquarters to unpaved Kumuwela Road branching *mauka* (to the south). Proceed 0.5 miles on Kumuwela Road to reach the trailhead and parking area for the Halemanu–Kōke'e and Kumuwela Trails, marked with a sign. Even if you don't hike the trails, I recommend stopping at this spot for a while. The towering koa trees overhead are breathtaking, and they are often frequented by forest birds. Go another 0.4 miles on the main dirt road, passing a branch road on the left, until you come to a fork in the road. A sign points the way to the trailhead for the Pu'u ka 'Ōhelo/Berry Flat Trail, up the hill to the left a short distance. You will see many private cabins near the trailheads, on land leased from the State.

To the right beyond this point the road is known as Camp 10–Mōhihi Road, and it has some sections that are often too slippery or muddy to drive in a 2WD vehicle. Even if the road is dry when you go in, a rain shower could make it too slippery to get out. I have seen foolhardy people in rental cars at the Alaka'i picnic area, nearly three miles from the pavement. I would never try it, and I strongly recommend against it. The road is deceptively smooth, but mud and lack of traction could leave you stranded. If you are in good condition, you can walk to the other birding trails from the Pu'u ka 'Ōhelo trailhead. There are some 4WD vehicles for rent on Kaua'i, but taking them off the pavement may violate the rental contract.

Once you make it to the Alaka'i picnic area, two miles past the Pu'u ka 'Ōhelo trailhead, you'll find a broad lawn area that looks very much

At the Alaka'i picnic area

out of place in the forest. There is a covered pavilion and picnic table at the far end. Even if you don't need to duck under the pavilion to get out of a passing shower, be sure to walk to the far end of the lawn to take in the view of the upper Waimea Canyon. There are good treetop views, too.

From the picnic area a spur road leads about a quarter mile to the trailhead for the Alaka'i Swamp Trail, while the main road continues about three-quarters of a mile to the trailhead for the Kawaikōī Stream Trail and beyond.

You can't miss the Kawaikōī trailhead; there's another large grassy picnic and camping area alongside the road just before the road fords Kawaikōī Stream. Trails follow both banks of the stream.

Halemanu–Kōke'e Trail This is a good, all-around trail: scenic views and canopy views, not too strenuous, well maintained, very little mud, and some good birding. 'Apapane, Kaua'i 'Amakihi, 'Anianiau, and 'Elepaio can be seen, as well as White-rumped Shama, Japanese White-eye, Hwamei, Red Junglefowl, Erckel's Francolin, House Finch, and Northern Cardinal. Tall grass along the trail catches mist and dew, so on a showery day or an early morning walk you may want rain pants even if it isn't raining. The trail extends for 1.2 miles before it intersects Halemanu Road.

Pu'u ka 'Ōhelo/Berry Flat Trail This loop trail provides a pleasant walk, including stretches that pass towering koa trees and groves of introduced redwood, sugi, and eucalyptus. Hurricane 'Iniki downed many large trees in this area, but the forest is still dense and shady. Thick understory sometimes obscures views of the birds, and there are only a few places where you have a view down on the forest canopy. Look for 'Elepaio, 'Apapane, Kaua'i 'Amakihi, Japanese White-eye, Nutmeg Mannikin, House Finch, and Hwamei. The trail passes through the forest for about 1.5 miles before it intercepts the dirt road again. Follow the road about 0.5 miles back to the trailhead.

Alaka'i Swamp Trail The Alaka'i Swamp Trail climbs from the Alaka'i Picnic Area until it reaches a broad, flat, boggy area. At this point the going gets very easy, because the next mile of trail is virtually all boardwalk. Topographically, this stretch of trail is not very interesting. The Alaka'i Trail intersects the Pihea Trail in a dense thicket. A little farther on, the Alaka'i Trail descends a steep slope and the boardwalk gives way to wooden stairs, which soon give way to a very steep, rocky, muddy trail. This difficult trail extends for two miles to a point called Kilohana, which is renowned for its views in clear weather.

Along the Alaka'i Swamp Trail

Instead of proceeding on this rugged trail, you might be better off sitting quietly where the trail starts to descend and waiting for the forest birds to come to you. Be sure to walk a short way south from the Alaka'i Swamp Trail on the Pihea Trail from the point where the two intersect. About a five- to ten-minute walk down the Pihea Trail is a rustic bench overlooking a deep canyon. It's a perfect spot for lunch or just quietly soaking up the sights, sounds, and smells of the forest.

The birds you are apt to see along the Alaka'i Swamp Trail are the same as the ones listed for the Pihea and Kawaikōī Stream Trails. In addition, be alert for the possibility of seeing some of the rarest birds on Kaua'i. It was along the Alaka'i Swamp Trail that a Nukupu'u may have been sighted in 1995.

Kawaikōī Stream Trail Many travel writers emphasize how beautiful this streamside trail is, and I have to agree. This area is about as close to most peoples' idea of an enchanted forest as you are likely to see. It isn't that the forest is pristine; there are quite a few introduced plants, including grasses, vines, and ginger. Part of the attraction may be that the ginger perfumes the air when it is blooming. And it isn't just the birds, although on my last trip to Kōke'e, Kawaikōī was the only place I saw an 'I'iwi. The scenic beauty of the stream has a lot to do with it, certainly: I can't think of a waterway anywhere that surpasses this stream for lush beauty. Maybe it's a combination of all these things. Whatever the reasons, Kawaikōī Stream is a place you are not likely to forget, and a trail you'll want to walk again. Don't be surprised if you are planning your next visit before you're even back to the highway.

There are trails on both sides of the stream leading upstream from the point where the road crosses the water. That's good, because during wet weather it may be too dangerous to ford the stream where it crosses the road, and you'll be restricted to the trail on the north side. Even in dry weather, you may want to take a small towel with you to dry your feet after wading across the ankle-deep water to reach the trail on the south side of the stream. You may not see the north-side trail on your maps. Sometimes only the south trail is shown.

To reach the trailhead for the north-side route, enter the grassy, fenced roadside picnic area with pavilions, a hundred yards before you reach the stream, and find the gateway through the fence at the back side near the stream. The trail parallels the water for about three-quarters of a mile, passing a suspended gondola used to cross the stream (except that it's locked, so you can't use it). Past a streamside pavilion the trail heads north, away from the water, and becomes the Pihea Trail.

The trail on the south side of Kawaikōī Stream is closer to the water and offers more views of the stream. If the stream is low enough to allow crossing where the road fords, do so and continue another hundred yards or so to the trailhead in the dense sugi grove. The trail backtracks, parallel to and uphill from the road, and then follows the stream for about three-quarters of a mile to a point near the pavilion on the north bank. There are well-positioned rocks near this spot that allow you to hop across the stream; do not cross if the water is high. Head back downstream on the north side trail, or continue north on the Pihea Trail.

Endemic birds along the stream might include 'Apapane, Common 'Amakihi, 'Elepaio, 'Anianiau, and 'I'iwi. Other bird species include Northern Cardinal, Hwamei, House Finch, Japanese White-eye, Common Myna, and White-rumped Shama. Spotted Doves may be heard fleeing forest perches noisily as you approach. Nutmeg Mannikins feeding on trailside grasses may be joined by a few Chestnut Mannikins. Hawaiian Coots can sometimes be seen on the stream, and Erckel's Francolins along the trail.

The Pihea Trail can be followed north for a little over a mile to the intersection with the Alaka'i Swamp Trail, or for about three miles to the Pu'u o Kila Lookout. Large trees toppled by Hurricane 'Iwa in 1982 are still visible along the trail, now joined by trees downed in 1992 by 'Iniki.

2. Kīlauea Point National Wildlife Refuge

FEATURES:
- Best marine bird viewing in Hawai'i
- Spinner dolphins and green sea turtles
- Very picturesque lighthouse

FACILITIES:

One of the premier birding spots in Hawai'i is the Kīlauea Point National Wildlife Refuge. This refuge has it all—a variety of marine birds that the public cannot observe elsewhere, great scenery, and the possibility of seeing marine life such as sea turtles, whales, and spinner dolphins. A whopping three hundred thousand visitors enjoy the refuge every year, making it one of the most popular wildlife refuges in the nation.

Kīlauea Point was a U.S. Coast Guard lighthouse from 1913 to 1976, when management was turned over to the U.S. Fish and Wildlife Service. It was formally designated a National Wildlife Refuge in 1985. In 1988 the acquisition of adjacent parcels of land helped to secure the seabird colony and provide a buffer. Today the refuge totals 203 acres in area. Offshore, the waters are protected as the Kīlauea Point National Marine Sanctuary.

The refuge was devastated by Hurricane 'Iniki. The visitor center was damaged, the shop where volunteers sold books and answered questions was totally destroyed, and native plants that had been reestablished and nurtured were swept away. Refuge fences blew down, allowing access to marauding dogs that killed defenseless Wedge-tailed Shearwaters in their burrows.

The birdlife at Kīlauea recovered quickly, but it took twenty months to get the refuge ready for human visitors again, and even then the services were limited. The visitor center had not reopened and volunteer-led birding walks had not resumed. By 1996 the refuge should resume its previous level of interpretive services. This will include volunteers assisting with the operation of the refuge, staffing a visitor center and bookshop, answering questions, aiming spotting scopes at nesting birds, lending binoculars to visitors, and leading nature walks several times each week. These two-hour walks follow a two-mile route up nearby Crater Hill, offering vistas not available from the peninsula where the lighthouse sits. These free walks are limited to fifteen participants, and space can be reserved by calling the refuge (see appendix).

A guided walk to Crater Hill

A Red-footed Booby chick

Which birds you see at Kīlauea will depend on the time of year that you visit. Table 6 below summarizes the seasonal occurrence of birds in the area. Red-footed Boobies are present year-round, and the thirteen hundred resident pairs breed nearly year-round, so there is always booby activity. Look for them on the hillside across the cove from the lighthouse to the east, where these birds nest in trees.

Brown Boobies do not nest at Kīlauea, but they are regularly seen in the area, especially in summer. Their nearest breeding grounds are on Lehua islet, just north of Ni'ihau, and Ka'ula Island, an isolated rock southwest of Ni'ihau.

About fifteen hundred pairs of Wedge-tailed Shearwaters nest in the area, arriving in March. These birds dig burrows—sometimes seven feet long—in which to nest. Many nest on Moku 'Ae'ae, an islet just offshore, but hundreds of pairs dig their burrows in the shrubbery surrounding the lighthouse where they can be approached quite closely. Look beneath the native beach naupaka that skirts the lawn area at the lighthouse for birds sitting at the entrance of their burrows. The lengths of plastic pipe visible atop Moku 'Ae'ae are artificial burrows placed there by the State of Hawai'i because the unstable soils on the island sometimes cause burrows to collapse. Biologists are experimenting with the burrows to see whether the shearwaters will use burrows narrow enough to exclude predatory feral cats that can foil nesting attempts on the main islands. The young Shearwaters fledge between mid-November and early December.

About the time the Shearwaters are leaving, the first Laysan Albatrosses are arriving, in middle to late November. These birds nest on the hillside to the west of the lighthouse. A grassy expanse is maintained near the nesting area as a landing site for these birds that are so graceful in the air and so clumsy on the ground. The last albatross chicks fledge in mid-July.

Birders are dwarfed by tree ferns at Thurston Lava Tube.

Hawai'i 'Amakihi in 'iliahi (sandalwood)
at Hosmer Grove.

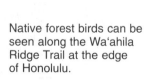

Native forest birds can be
seen along the Wa'ahila
Ridge Trail at the edge
of Honolulu.

The Nēnē can be seen on Kauaʻi, the Big Island, and Maui.

Visitors such as this Northern Cardinal
are easily attracted with bird seed.

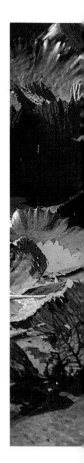

Zebra Doves, common in urban and residential areas,
are remarkably tame.

Common Moorhens in parks can be approached quite closely.

Kawaikōī Stream on Kauaʻi offers good birding and good scenery.

Red-footed Boobies nest at Kāneʻohe on Oʻahu and at Kīlauea Point National Wildlife Refuge on Kauaʻi.

Park Service rangers lead hikes in the Waikamoi Preserve on Maui.

Great Frigatebirds, like this juvenile, are frequently seen in flight.

Laysan Albatross can be seen at Kīlauea Point on
Kaua'i and occasionally at Ka'ena Point on O'ahu.

A Wedge-tailed
Shearwater
in its burrow at
Kīlauea Point.

A Black-crowned Night-Heron along
Kaʻelepulu Stream on Oʻahu.

The flower of the ʻōhiʻa, called lehua, provides nectar
for birds and makes the tree easy to recognize.

The native 'ōhi'a tree
is very important to
Hawaiian forest birds.

The koa bears leaf-like phyllodes or
flattened stems as a mature tree.

The fruits of the native naio, or false sandalwood,
are eaten by birds in dry areas.

The green seed pods
on this māmane tree
at Puʻu Lāʻau are an
important food source
for the Palila on the
Big Island.

Table 6
Seasonal Occurrence of Birds at Kīlauea Point

BIRD	J	F	M	A	M	J	J	A	S	O	N	D
Red-footed Booby			———	——	——	——	——	——	——	——	———	
Brown Booby						——	——	——				
Wedge-tailed Shearwater					———	——	——	——	——	——	———	
Laysan Albatross	———	——	——	——	——	——	——	——			———	——
White-tailed Tropicbird	———	——	——	——	——	——	——	——	——	——	——	——
Red-tailed Tropicbird			———	——	——	——	——	——	——			
Great Frigatebird		———	——	——	——	——	——	——	——	——	——	——
Kōlea	———	——	——	–	–	–	–	–	–	———	——	——

Both the Red-tailed Tropicbird and the White-tailed Tropicbird breed at Kīlauea. Their breeding season starts in March, and the young fledge between June and September. The White-tailed Tropicbird can be seen in the area year-round, but the Red-tails disperse after their breeding season and are not as likely to be seen at the refuge after September. The best time to see these birds performing the aerial acrobatics of their courtship is in March and April.

A prominent resident of the refuge that is seen year-round is the Great Frigatebird. These birds are not known to nest on Kaua'i, and most of the frigatebirds seen at Kīlauea are females that have probably dispersed southward from the Northwestern Hawaiian Islands. A few males have been seen in recent years, one sporting a partially inflated red throat pouch indicative of a breeding male. Perhaps in time, this bird will begin to nest at Kīlauea.

Newell's Shearwaters do not nest at Kīlauea, preferring fern-covered slopes up in the mountains. Sometimes sharp-eyed observers spot them in the waters off the point. See additional information below.

There are several other noteworthy birds to watch for. The Kōlea winters at Kīlauea and can often be seen around grassy areas and roadsides from August through April. A few remain all year. Visitors on the guided walk may spot Western Meadowlarks in the hillside meadow. Finally, keep an eye out for Nēnē. Since 1991, thirty-one of these birds have been released on Crater Hill in hopes that they would establish a breeding colony in the protection of the refuge. Despite some early casualties, the flock produced several goslings late in 1992, after Hurricane 'Iniki, and has been breeding successfully ever since.

Lovers of nature will find that birdlife is not the only attraction at Kīlauea. The Fish and Wildlife Service is working to restore native plants to the area, and several are identified with excellent interpretive signs. Unfortunately, it will take years to restore the damage done to native plants by Hurricane 'Iniki. Green sea turtles are often seen in the waters around the peninsula, as are pods of Pacific spinner dolphins. From December to May, visitors may spot humpback whales that winter in these waters. On rare occasions the endangered Hawaiian monk seal is spotted in the summer.

The refuge is open from 10:00 A.M. to 4:00 P.M. weekdays, and closed federal holidays. A modest admission fee is charged. There is very good wheelchair access to rest rooms, drinking fountain, visitors center, and bird viewing areas. There is an electric cart available for the use of visitors who cannot walk from the parking lot to the lighthouse on the paved trail.

A Seabird of the Mountains

One of Kaua'i's most intriguing resident seabirds doesn't nest at Kīlauea or anywhere else right along the coast. Newell's Shearwaters make their nests on fern-covered slopes in the island's mountainous interior. Like the Wedge-tailed Shearwaters at Kīlauea, they dig burrows in which to rear their young. These birds are seldom seen: from November to April they apparently leave Hawaiian waters completely, and during their breeding season they come and go from their nest burrows only at night. Breeding begins in April, and the young birds fledge in October and early November.

The first flight of the fledglings is a perilous one. They apparently cannot take wing from level ground—only from their burrow entrances on steep mountain slopes. During their first nocturnal flight, they seem to head instinctively toward light. Moonlight reflected on the ocean is the perfect natural beacon to guide them to the sea, where they will spend the first years of their lives.

As the human population of Kaua'i has increased, so has the artificial lighting of resorts, parking lots, athletic fields, and city streets. When the young shearwaters make their first flight to the sea each autumn, they can become confused by these lights and fly into buildings, cars, trees, and utility wires. The problem is most severe at new moon, or on overcast nights when the moon isn't visible. The downed birds are often uninjured, but these seabirds are adapted to take flight from water and cannot take off from level ground. They are at the mercy of dogs, cats, and traffic.

In the 1950s only a few downed birds were noticed each year. In 1967, two hundred birds were downed on the grounds of a single new resort in Līhu'e. Now there are so many lights to distract the birds that about fourteen hundred are rescued each year by helpful Kaua'i residents and visitors. Shearwater aid stations are set up at every firehouse on the island, where downed birds can be deposited in cages for pickup by government biologists. Most birds are retrieved in the populous and illuminated areas of Līhu'e, Kapa'a, and Po'ipū. The main fledging and rescue period runs from early October through the first week in November, peaking about October 25.

In early September 1992 when 'Iniki hit, young Newell's Shearwater chicks were helpless. Many may have been buried or drowned in their burrows as the storm swept across the island. An indirect effect of the hurricane, loss of habitat, may be even more devastating to the shearwater populations than the direct effects of wind and rain. When adult birds return to the island in future years, they may find old burrows destroyed and familiar fern-covered slopes transformed by the addition of alien plants. Only time will tell what the full impact of Hurricane 'Iniki will be on these birds.

From late April to early October, adult birds can sometimes be seen near dusk as they gather offshore before heading inland to their colonies. The best places to look are offshore from the areas where most downed fledglings are found, such as Kapa'a and Po'ipū.

If you visit Kaua'i in the fall and you are out early in the morning, you may find a hapless Newell's Shearwater on the roadside where it has become confused by city lights. Don't be intimidated by the bird's thirty-two-inch wingspan; it will most likely be dazed and docile. Take your feathered friend to the nearest firehouse so this threatened species can continue to coexist with the more recent human inhabitants of its island.

3. Hanalei National Wildlife Refuge

FEATURES:
• Endangered wetland birds
• Taro patches

FACILITIES:

Hanalei Valley

One of the most picturesque scenes in all Hawai'i is the view of Hanalei National Wildlife Refuge from the overlook on Highway 560, near the town of Hanalei on Kaua'i's north shore. The refuge does double duty as a wildlife haven and a productive agricultural area. It is a patchwork of flooded fields where taro is grown along the banks of the Hanalei River. The cultivation of taro in these wet plots provides a nearly ideal habitat for endangered Hawaiian wetland birds: Hawaiian Coot, Common Moorhen, and Black-necked Stilt. It is one of those happy situations where everyone wins: taro farmers cultivate the rich bottomlands of the Hanalei, and the rest of us have the perfect place to observe wetland birds.

The highway overlook is just past the Princeville Shopping Center, at the point where Highway 56 becomes Highway 560. After you have taken in this expansive view, continue down the road for 1.1 miles. Just past a narrow bridge that takes you over the Hanalei River, narrow but paved 'Ōhiki Road leads off from your left up the river past the taro patches of the refuge. This is a public road and you can drive it, but keep in mind that you will be sharing the narrow road with the taro farmers who live and work in the area.

Although a car makes a good blind to obscure your silhouette from the birds, you may choose instead to park your car at the large turnout on your left just before you cross the bridge and walk along 'Ōhiki Road.

You'll be able to stay out of the way of the farmers, and you may see more birds. The road is level and you can walk about a mile and a half up the valley, but the best birding is near the foot of the road where it is bordered closely by taro patches.

Watch for Black-necked Stilts standing on the dikes between the taro patches. Hawaiian Coots and Common Moorhens can be spotted along the dikes, in the water-filled patches, or down in the Hanalei River. If you are lucky you may see Koloa in the patches or along the river. Also watch for Black-crowned Night-Herons in the area. Cattle Egret is another species almost always found. Where trees and shrubbery interrupt the taro patches, look for Hwamei, White-rumped Shama, Japanese White-eye, Japanese Bush-Warbler, and Northern Cardinal. The Fulvous Whistling-Duck has also been spotted.

At times the birds will be too distant for a good look, or the high dikes along 'Ōhiki Road may partially obscure your view. Numerous signs along the way will inform you that all the refuge land off the road is closed. Straying from the road may alarm the birds and damage the fragile dikes between the taro patches. It will also earn you a ticket if any refuge employees see you.

If you walk, take along drinking water, a sun hat and sunscreen to protect you on hot sunny stretches of the road, and rainwear for those frequent passing showers that blow through. There are no rest rooms in the area. The nearest ones are back at the Princeville Shopping Center, where access for the disabled is good, or ahead at Hanalei Beach County Park.

Birding by Boat

A very different perspective for viewing wetland birds is from the water. Several rivers and streams on Kaua'i are navigable by canoe or ocean kayak, including the Hanalei River, the Wailua River, the Kalihiwai River, and Hule'ia Stream. Among these, the best birding waters are the Hanalei and the Kalihiwai. There is too much boat traffic on the Wailua, and both it and Hule'ia Stream are heavily overgrown with hau trees, lacking the grassy banks preferred by water birds.

From a boat you may not see more birds than you can from land, but you will get to see them from their own perspective on the water. Moreover, it is usually possible to approach the birds more closely in a quiet boat than you can by land. Remember that the birds, especially Hawai'i's endangered wetland birds, should not be harassed in any way.

There are several firms around the island that rent boats for river

On the Hanalei River

exploring. You can usually find their brochures in the racks of tourist literature at the Kaua'i airport and other locations. Different types of boats are available, including rigid canoes, inflatable canoes, and ocean kayaks. Rental includes paddles, life vests, and equipment to carry the boat on your car from store to river.

A four-mile stretch of the Hanalei River can be paddled from the mouth at Hanalei Bay Beach Park up through the Hanalei National Wildlife Refuge. The first two miles of the river, from the mouth to the highway bridge, are below the refuge. Although the river is lined with houses for part of this stretch, and follows the road for another part, this is still a good area to see birds. Watch especially along grassy banks for Hawaiian Coot, Common Moorhen, and Koloa. You may see Black-crowned Night-Herons perching in the trees along the bank.

The trees with nearly round leaves that grow in great thickets close to the water's edge are hau. The hau flower lasts only about 24 hours, changing color from yellow, to orange, and then to red before dropping from the plant. While passing these trees watch and listen for Japanese White-eye, Hwamei, White-rumped Shama, Northern Cardinal, Japanese Bush-Warbler, and Common Myna.

Paddling farther upstream yields more of the same birds. There are neither more birds nor more bird species once you have entered the wildlife refuge. You probably will not see the Black-necked Stilt along the river. This bird prefers the taro patches, which are obscured from view by the high banks of the river.

During rainy periods the Hanalei may be running too high to paddle, and the boat rental shops will probably not rent to you. At other times there is practically no current at all. As you paddle up the river, remember that you will also have to paddle back down since the current is too slow to carry you.

Rest rooms and drinking water are available at Hanalei Bay Beach Park. Once you're upstream of the park you're on your own: there is no place to get out of your boat except in the water because the shore is private land or closed refuge land. You will be in the boat for several hours. Be sure to take water and perhaps a lunch. Waterproof sunscreen is another necessity, because much of the river is in the sun.

4. Wailua River

FEATURES:
- A grab-bag of birds
- Splendid views of river and falls
- A *heiau*

FACILITIES:

The waters of the Wailua River have carved a broad, deep valley into eastern Kaua'i. This valley between Līhu'e and Kapa'a offers many sightseeing opportunities, including two botanic gardens, a spectacular waterfall, an ancient *heiau*, and hiking trails. Every attraction boasts some notable birding.

Starting at the mouth of the river and moving *mauka*, the first stop is an attraction called Smith's Tropical Paradise, adjacent to the marina on the south side of the river. This tidy thirty-acre botanic garden might be a little too commercial for most birders, if it were not for the good birding. For starters, there are usually Peafowl strutting around the parking lot. The lily ponds in the garden are home to quite a few Common Moorhens. This species is usually rather wary, but these birds can be approached closely enough to get good photographs with a short tele-photo lens. Fulvous Whistling-Ducks can also be seen near the ponds. In the dense shrubbery, watch for Greater Necklaced Laughing-thrushes. There is an admission fee. Hours are 8:30 A.M. to 4:00 P.M.

The rest of the birding spots along the Wailua River are all on the north side, along Highway 580. To reach them, follow Highway 580 *mauka* from its junction with Highway 56 in Wailua. At the 1.4-mile point on Highway 580, a one-way circular driveway to the left leads to the Poli'ahu Heiau overlooking the Wailua River. White-tailed Tropicbirds may be seen soaring over the broad valley cut by the Wailua. The cattle grazing in the bottomlands far below you will almost certainly be accom-panied by a faithful cadre of Cattle Egrets. In the lawn around the *heiau* look for Nutmeg Mannikins and smaller numbers of the darker Chestnut Mannikin.

Just up the road on the right is the turnout for 'Ōpaeka'a Falls. An overlook provides a good view of the gulch and cliffs cut by 'Ōpaeka'a Stream. Watch for White-tailed Tropicbirds soaring over the gulch. The feral chickens in this area are descended from the Red Junglefowl brought to Hawai'i in ancient times by the Polynesians, but these birds have interbred with domestic chickens and are not pure Junglefowl. Other birds to watch for include Common Myna, Zebra Dove, Spotted Dove, Kōlea, Northern Cardinal, Red-crested Cardinal, and Cattle Egret.

Farther up Highway 580 is Wailua Reservoir. It is not open to the public but there is good viewing from the entrance roads. A road at the five-mile point leads to the lower part of the reservoir, where you may see a few Hawaiian Coots. A little farther up, the highway passes over a stream that feeds into the reservoir. This is another place to look for coots. Just above the stream is a private dirt road. From the gate you will have a good view of the upper part of the reservoir.

Still farther up the road is the trailhead for the Kuilau Ridge Trail, at the 6.7-mile point. This trail climbs for 1.3 miles to a grassy area with pic-nic pavilion and superb views. You probably won't encounter any avian rarities along this trail, but if you hike it for the views and the exercise, look for Hwamei, White-rumped Shama, Japanese White-eye, and Com-mon Myna. You may even flush a Ring-necked Pheasant from the trail-side. Expect a little mud along the trail, and carry mosquito repellent.

If you drive a few hundred yards past the trailhead for the Kuilau Ridge Trail, you will come to Keahua Arboretum. This is usually not a great birding spot, but I have seen Greater Necklaced Laughing-thrushes in this area. These large birds are usually easy to spot if they are present.

Look for them in small flocks near the top of the dense streamside vegetation. The stream that crosses the road and bisects the arboretum is often too high to cross at the ford. Park before crossing the stream and walk downstream a few hundred yards. This is usually a good, quiet spot for a picnic, and better for birders than the arboretum land across the stream. Beyond the arboretum, the road quality goes downhill fast, so there is really no point in trying to cross the swift-moving water.

5. Mānā Ponds

FEATURES:
• Endangered wetland birds

FACILITIES: None

The lowlands at the west end of Kaua'i, from Kekaha out to the Barking Sands Missile Range, were once vast wetlands that supported thousands of resident birds and migrant waterfowl. Long ago this area, called the Mānā Plain, was drained and planted in sugarcane, but there are still places to see wetland birds in the area.

As you drive along Highway 50 past Kekaha, you will notice deep ditches cutting across the flat terrain. Most of these ditches are difficult to explore, but many of them harbor Koloa, Hawaiian Coot, and Common Moorhen. If you can pull off the highway to check one of these ditches, you might be rewarded with a glimpse of some of the state's endangered waterfowl.

A more accessible place to see birds in the area is at the very end of Highway 50. At this point, just past the thirty-two-mile marker, the designated State highway ends and the road splits. The right branch leads out to Polihale State Park, while the left branch dead-ends in one mile, at a gate for the Pacific Missile Range. On your left, 0.6 miles down this left branch, is a pond area that harbors a good variety of wetland birds. The Black-necked Stilt nests at the pond, and can almost always be seen in the shallow, open pond area. To view the pond you must scramble up a low embankment next to the road. The stilts are very wary, and will often call continuously as long as you are visible above the embankment. As soon as you go back down to the road out of their sight, they will stop.

Just past the pond, at the very end of the paved road, a dirt road to the left leads you a few yards off the pavement to spots where you can look down into deep ditches. Look for coots and moorhens in the water. They will often keep to the vegetation at the sides of the ditches, venturing into open water only when they want to make their way along the ditch. Black-crowned Night-Herons can sometimes be seen in the trees along these ditches, and may make a noisy exit as you approach.

The ditches and ponds make the area attractive for migratory waterfowl. In addition to the resident Koloa, you may spot some migrants in the winter, such as Northern Pintail or a less common visitor. Other resident birds to watch for include Northern Cardinal, White-rumped Shama, Spotted Dove, Common Myna, Japanese White-eye, and Cattle Egret.

6. Kōloa Area

FEATURES:
• Convenient birding for Po'ipū visitors
• Chance to see Greater Necklaced Laughing-thrush

FACILITIES:

Many visitors to Kaua'i stay at hotels or condominiums in the Po'ipū Beach area just down the road from the old sugar town of Kōloa. If you find yourself in the area, there are several birding spots you will want to know about.

Just outside picturesque Kōloa Town is a quiet country road that offers a good spot to stop and do some birding. From Kōloa go west on Highway 530 for 0.4 miles to 'Ōma'o Road. Drive up 'Ōma'o Road another 0.4 miles and park at the turnout just past the narrow bridge. A few feet off the road, and nearly obscured by vines and hau trees, is tiny Loko Reservoir along 'Ōma'o Stream. During a recent visit I found the reservoir drained, but if it is full you may spot Hawaiian Coots enjoying this small lush pool. Stand quietly for a few minutes and you are likely to see or hear White-rumped Shamas in the trees. If you are lucky you may spot a Hwamei (sometimes known as Melodious Laughing-thrush) or the less common and larger Greater Necklaced Laughing-thrush. Across the highway in the open area by the narrow bridge, look for Cattle Egret,

Zebra Dove, Spotted Dove, Japanese White-eye, Nutmeg Mannikin, Chestnut Mannikin, House Finch, Red Junglefowl or feral chicken, and Common Myna. Less common birds to watch for include Western Meadowlark, Northern Mockingbird, and Warbling Silverbill.

A famed attraction in this area is the "tunnel of trees" along Maluhia Road or Highway 520 between Highway 50 and Kōloa, where rows of towering swamp mahogany arch over the road. As you drive this road you can see Mauka Reservoir partially obscured by the understory to the west of the road. It is difficult to stop along this stretch of road because most of the turnouts are really too small for a car and it is tricky merging back into the fast-moving traffic. The bird most likely to be seen on Mauka Reservoir is the Hawaiian Coot, sometimes dozens of them, but they can be seen elsewhere.

Just north of Kōloa is Waita Reservoir, the largest body of water on the island. Migrant waterfowl can sometimes be seen at Waita in the winter. The reservoir is on private sugar land. To visit, you must get permission from the McBryde Sugar Company. Their office and sugar mill are in the town of Numila, near Hanapēpē.

Condominiums surrounded by landscaping at Po'ipū are good places to attract birds to your lanai or patio. Expect to see House Sparrow, Zebra Dove, Northern Cardinal, and Red-crested Cardinal. Common Mynas usually disdain birdseed, but will be attracted by junk food such as bread or crackers. Birds that will not be attracted by the food you set out, but may approach out of curiosity, are White-rumped Shama and Japanese White-eye.

Along grassy roadsides in this area you may see Nutmeg Mannikins and Chestnut Mannikins, often hanging on tall grass stalks. These birds, along with Zebra Dove, Spotted Dove, Common Myna, Kōlea, and Cattle Egret can be seen in lawn areas. A good place to see Cattle Egrets is in the groundcover along Po'ipū Road near the hotels and condominiums. In the trees around hotels or in dry scrub areas, look for Northern Cardinals and Red-crested Cardinals.

Around ponds at golf courses and landscapes, you may see Black-crowned Night-Herons doing a little fishing. Occasionally a Great Frigatebird will glide high overhead. At dusk large flocks of Cattle Egrets may gather and fly over Kōloa toward their roosts.

7. Rainy Day Birding—Kaua'i Museum

FEATURES:
- Rare Hawaiian feather work
- Good historical and cultural displays

FACILITIES:

A rainy day birding option on Kaua'i is the Kaua'i Museum in Līhu'e. This small museum concentrates on the people and history of Kaua'i. A few natural history displays are being developed. Birders will marvel at the few pieces of Hawaiian feather work in the museum collection: leis of 'Ō'ū and Mamo feathers, and capes of 'I'iwi and 'Ō'ō feathers.

The museum is open Monday through Friday from 9:00 A.M. to 4:30 P.M., Saturday 9:00 to 1:00, and closed Sundays and holidays. An admission fee is charged. There is wheelchair access to the old stone structure that houses the museum through the office at the back of the building.

Other Kaua'i Birding Spots

Kē'ē Beach

If you follow the highway for seven twisting, turning miles past Hanalei on Kaua'i's north shore, you come to road's end at Kē'ē Beach in Hā'ena State Beach Park. This spot boasts of great snorkeling in calm weather and is the trailhead for the famous Kalalau Trail, which snakes along the cliffs for another eleven miles. The scenic grandeur of the route is legendary; this trail often appears on lists of the best backpacking trips in the world.

If you come for the hiking or the snorkeling or the views, be sure to gaze out over the water. Kē'ē Beach seems to be a particularly good spot for seabird activity. Watch for Brown Booby, Great Frigatebird, and White-tailed Tropicbird. Only the tropicbirds nest on Kaua'i. Brown Boobies nest on Lehua islet, just north of Ni'ihau, and on remote Ka'ula Island, beyond Ni'ihau. Frigatebirds are visitors from the Northwestern Hawaiian Islands.

Frigatebirds have found a way to feed without diving for their dinner. They sometimes harass boobies and other birds in flight until the victims disgorge their food. This practice led early sailors to call the Great Frigatebird the Man-o'-War Bird after the term used for armed naval vessels. Watch for this aerial show at Kē'ē Beach.

Kaua'i Marriott Resort

The Kaua'i Marriott Resort in Līhu'e is situated on a large site that includes extensive landscaped grounds, lagoons, and islands. The park-like setting is home to domestic ducks and swans. The lush grounds, the water, and the domestic birds attract many wild birds as well. Some of the species on the hotel grounds include Spotted Dove, Zebra Dove, Common Myna, Northern Cardinal, Japanese White-eye, House Sparrow, House Finch, Nutmeg Mannikin, Chestnut Mannikin, Cattle Egret, Hawaiian Coot, Common Moorhen, Koloa, Kōlea, White-rumped Shama, and Java Sparrow.

The big surprise is a flock of over a hundred wild Nēnē that occasionally visit the area, settling in on an island in the lagoon. These birds are descended from a few caged birds that escaped from their pens in 1982 during Hurricane 'Iwa. Nobody told these endemic geese that they were supposed to be endangered, or that scientists are trying to establish them in higher-altitude upland areas of Maui and the Big Island. So they have been successfully breeding in the lowlands of Kaua'i, on private land a stone's throw from the ocean, and occasionally showing up at the Marriott for lunch.

Boat tours featuring the wildlife are available for a fee, every hour from 9:00 A.M. to 5:00 P.M. daily. The hotel is right in Līhu'e, near the airport and the harbor. Enter the resort grounds from either the Kapule Highway (Highway 51) or the foot of Rice Street. Attendants at the gate will direct you to the visitor center, where you can board a boat to see the Nēnē.

Kukui o Lono Park

This is a place where you can do some birding in a nicely maintained town park, see the usual variety of urban birds, and maybe spot Rose-ringed Parakeets. To reach the area, take Highway 50 to the town of Kalāheo. Head south on Papalina Road, which intersects the highway at the only traffic light in town. The entrance to the park is one mile south of the highway.

Admission to the park is free, and it is open daily from 6:30 A.M. to

6:30 P.M. Much of the park is devoted to a golf course, where Red Jungle-fowl or feral chicken can be seen, along with Kōlea, Common Myna, Nutmeg Mannikin, and Chestnut Mannikin. In the trees near the fairways look for Northern Cardinals and Red-crested Cardinals. Just inside the front gate a nature trail loop takes off to the right through a thick grove of trees. This trail was closed after the hurricane because of downed limbs, but it may be open again by the time you visit. It is a good area to see White-rumped Shamas. The lucky visitor to Kukui o Lono may see Rose-ringed Parakeets. Flocks of these birds occasionally roost in the trees at the top of the hill in the park. Watch for them coming in at dusk.

Salt Pond Beach County Park

Another birding spot that is worth a stop if you are nearby, Salt Pond is at the west end of Hanapēpē, just off Highway 50. To get to the park, go south from Highway 50 on the well-marked entrance road and follow the signs to the park.

Adjacent to the park are two evaporation ponds that have been used to produce salt for hundreds of years. In the summer and fall these ponds are filled with evaporating brine. In winter and spring freshwater runoff fills them, providing habitat for wetland birds. The Black-necked Stilt is the most common visitor; other birds you may see in this area include Cattle Egret, Black-crowned Night-Heron, Northern Cardinal, Red-crested Cardinal, Northern Mockingbird, Kōlea, Spotted Dove, Zebra Dove, Common Myna, and House Sparrow. On the beach look for Wandering Tattlers.

6
The Big Island

IMAGINE A GROUP of happy birders, hiking through misty woodlands, surrounded by snowy mountain peaks, spotting rare and endangered birds, marveling at pockets of native forest, without another person for miles. Does this sound like it could be Tibet? Now imagine these happy birders continuing their day's activities with a visit to an active volcano where they see more endemic forest birds, and topping things off with some serious beach basking surrounded by colorful exotic birds, a little snorkeling and perhaps a swim with sea turtles. The Big Island may be the only place on earth where you can treat yourself to this kind of sensory overload. For people with a broad interest in nature, this is the island of choice.

You certainly won't feel confined on the Big Island, with over four thousand square miles to explore. This island claims almost two-thirds of the total land mass of the state, and the total is growing as Madame Pele continues to add to the island's size with lava flows that extend the shoreline. All the space means you'll have to allow a little more driving time on the Big Island. A one-day drive around the island is possible, but it would leave precious little time for birding or sightseeing.

The Big Island can also lay claim to being by far the highest in the Hawaiian chain, with Mauna Kea at 13,796 feet and Mauna Loa at 13,679. Next to these giants, 8,271-foot Hualālai on the island's west side goes almost unnoticed. The high elevations are very important to birds. A huge area is safely above the range of mosquitoes (though you'll also find some Big Island endemic birds coexisting nicely with mosquitoes at lower elevations). The uplands are incredibly diverse, ranging from lush rain forest on the windward face of Mauna Kea to moonscapes at the mountain summits to grasslands and dry woodland on the high slopes and leeward side of the island.

The high-elevation areas of the Big Island are very sparsely populated. Most residents and visitors stay near the coastline. The dry west coast from Kailua-Kona north to Waikoloa draws most of the visitors, but the sunny resorts on this side of the island are a long way from the best places to see forest birds. Hilo, on the east side, is much closer to good mountain birding areas but is less popular as a vacation spot because of the frequent showers on this windward side of the island.

Other lodging choices put you very close to some of the island's

Map 4
Big Island Birding Sites

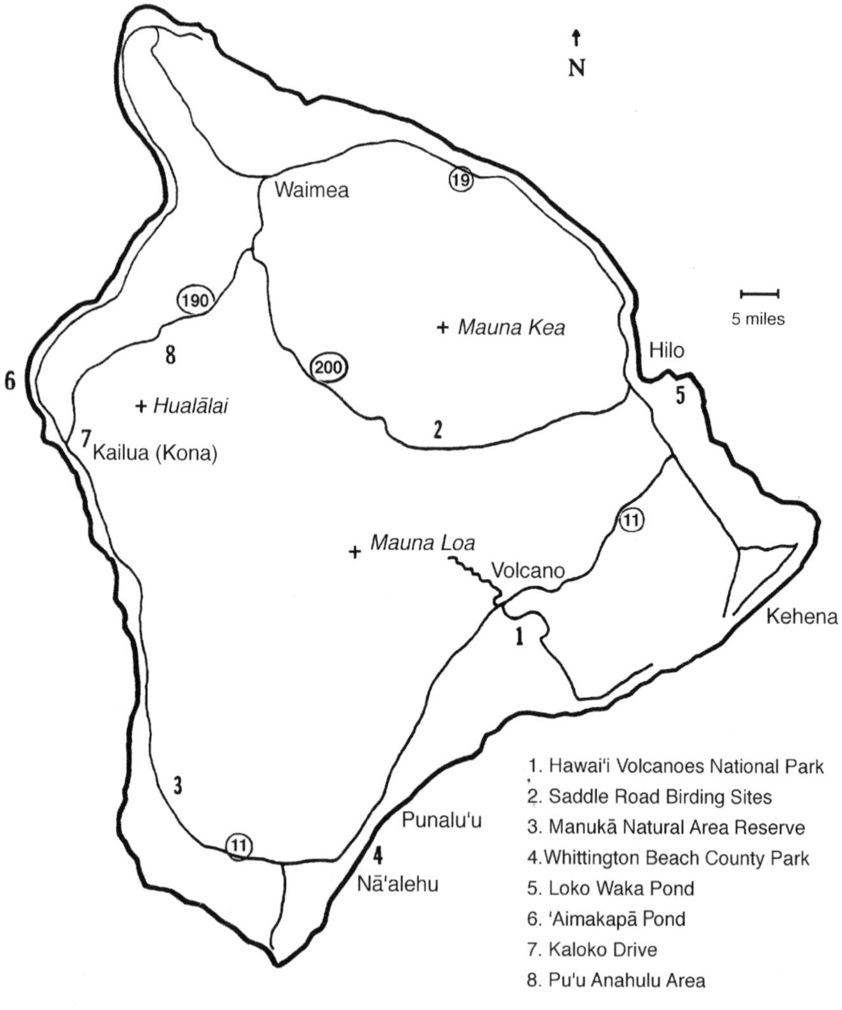
N
5 miles
Waimea
19
190
+ Mauna Kea
Hilo
200
8
6
5
+ Hualālai
2
7
Kailua (Kona)
11
+ Mauna Loa
Volcano
Kehena
1
3
Punalu'u
11
4
Nā'alehu
1. Hawai'i Volcanoes National Park
2. Saddle Road Birding Sites
3. Manukā Natural Area Reserve
4. Whittington Beach County Park
5. Loko Waka Pond
6. 'Aimakapā Pond
7. Kaloko Drive
8. Pu'u Anahulu Area

endemic birds. In the volcano area, Hawai'i Volcanoes National Park (HVNP) has an adequate hotel right in the midst of good birding areas, while the nearby town of Volcano has a few bed-and-breakfast establishments. The State of Hawai'i also maintains nice rental cabins at Pōhakuloa, along the Saddle Road between Mauna Kea and Mauna Loa. Many good birding spots are along this road. Birders who are drawn to the sunny warmth and luxury of Kona or Waikoloa may want to spend a night in a cabin at Pōhakuloa and a night at the volcano while birding in these areas.

The Big Island lays claim to over 70 percent of the public hunting areas in the state. This is good for birders, because many of these public areas are also hot birding spots. However, sharing the land with hunters requires some special precautions. The State Division of Forestry and Wildlife requires hikers and birders to obtain permits before entering hunting areas. The permit allows you to enter a particular area on specific days. It also warns you that hunting may be in progress and advises or requires bright-colored clothing. If you don't have an orange hunting vest, a pink or orange neon-colored tee shirt does the trick. All the birding spots that require a permit are noted in the text.

Permits are free and can be obtained by visiting the Division of Forestry and Wildlife Baseyard in Hilo. The baseyard is open weekdays except State holidays from 7:45 A.M. to 4:30 P.M. Permits are also available at the State Tree Nursery in Kamuela (also known as Waimea). The nursery is located between the Kamuela airport and town on the west side of the highway. Hours are 7:00 A.M. to 3:30 P.M. (see appendix for additional information). It's a good idea to get your permit a day or two before you use it so you can get an earlier start when you visit an area. Most hunting occurs on weekends, so try to visit hunting areas on weekdays when you are more likely to have them all to yourself.

1. Hawai'i Volcanoes National Park

FEATURES:
• Volcanoes
• Good and varied birding

FACILITIES:

Sometimes nature is very kind to birders, giving us wonderful birds to see and wonderful surroundings in which to see them. Hawai'i Volcanoes National Park is such a place. The settings in the park range from lush rain forest to dry open woodland, from stark volcanic craters to sea cliffs where the waves have carved stone arches and transformed cooling lava into black sand. While you are marveling at these settings, you'll be surrounded by good birding. There are endemic forest birds, Nēnē, tropicbirds, a host of introduced birds, and marine birds to observe as you're taking in the other features of the park. This part of the Big Island can be crowded, but your birding will take you to the least-visited places in the park. As with so many other birding sites in Hawai'i, you'll have paradise almost all to yourself.

The visitor center at park headquarters is a good place to start. The center features impressive films of volcanic eruptions, displays on the plants and animals of the park, and a good selection of natural history publications. The rangers at this busy facility answer hundreds of questions on volcanic activity every day. Your questions about birds will provide them with a delightful change, so be sure to ask! The visitor center is also the place to find out about nature walks, lectures, and other activities that are held nearly every day. Look for information posted on the bulletin board by the front door.

From the visitor center you can take a short walk to the Volcano Art Center, housed in a rustic old building that served as a hotel many years ago. This gallery offers a delightful selection of local arts and crafts. Excellent prints and paintings of Hawaiian birds are usually available. The trees you pass as you walk from the visitor center to the art center are 'ōhi'a, a favorite food source for the 'Apapane. These endemic honeycreepers are often feeding in the area, so be sure to watch for them.

Some parts of the park can be quite rainy, so you may need a wet-weather strategy to get the most out of your visit. The coastal sections of the park receive far less rainfall than the area around the visitor center and Kīlauea Caldera. If you encounter rain up above, head down below to look for Nēnē and seabirds, visit some of the newest black sand beaches in the world, or marvel at petroglyphs. This will take at least a few hours, and by then the rain may have passed.

Kīpuka Puaulu (Bird Park)

Kīpuka is the Hawaiian word for a patch of vegetation that is surrounded by newer lava flows that have less, or no, plant life. *Kīpuka* are often called islands of vegetation. Kīpuka Puaulu is an area of old, dense forest that is surrounded by scattered younger trees, scrub, and lava flows.

The attractions of this area include a good loop trail that is perfect for an easy stroll, an interpretive brochure keyed to numbered points along the trail, and amazing solitude. The throngs of tourists visiting Hawai'i Volcanoes National Park each year are more interested in geologic than biologic features. Another reason for the area's solitude is that it is separated by Highway 11 from the other attractions of the park. You will usually have the trail and the birds all to yourself, especially early in the morning.

The area is also called Bird Park, a name that is enough to tantalize any birder. In a state blessed with wondrous birdlife, this place must be fabulous, right? Well, the birding can be very pleasant, but perhaps not as good as other areas described in this book. You have a fair chance of seeing four endemic species: 'Ōma'o are quite common, 'Elepaio fairly common, 'Apapane variable but often common, and Hawai'i 'Amakihi are sometimes seen. 'I'iwi are more common a little higher upslope on Mauna Loa, but may be seen at Kīpuka Puaulu occasionally. The drawback in this forested birding spot is that all these birds except the 'Elepaio dwell mostly in the canopy, high above your head. There is no place at Kīpuka Puaulu where you have good canopy views.

Introduced bird species are far more abundant than endemics in this *kīpuka*. Expect to see swarms of Japanese White-eyes. Northern Cardinals and House Finches are common, and the Red-billed Leiothrix and Hwamei may be seen. Also watch for Kalij Pheasants. Their numbers are increasing sharply in the area. If you approach quietly, you may be able to get within a few feet of these big birds.

To reach Kīpuka Puaulu, take Highway 11 west from the main park entrance to Mauna Loa Road and follow the signs. A well-maintained 1.2-mile loop trail takes you through this nearly flat area. Access to the area is outside the entrance gate to HVNP, so there is no admission charge. Kīpuka Puaulu has no wheelchair access beyond the parking lot on the edge of the *kīpuka*. There is also a fine picnic area along Mauna Loa Road just before you reach the trailhead, where you will find tables, pavilions, toilets, and drinking water surrounded by koa trees.

The Hawai'i Natural History Association has produced a nature guide to the trail. Copies to borrow may be available in a box at the trailhead; to be safe purchase one for $0.50 at the park visitor center before coming to Kīpuka Puaulu. The informative five-page guide, keyed to numbered points along the trail, gives a good background on *kīpuka* and the effects of introduced species.

Another nearby birding spot worth mention is the Volcano Golf Course, north of Highway 11 between the park entrance and Mauna Loa Road. This is a fairly dependable place to see Nēnē, although certainly not a very natural setting in which to view these mostly banded and captive-reared birds. If you don't see Nēnē around the clubhouse, try walking along the gravel road past the helicopter concession office at the southern boundary of the course. The birds sometimes hang out at the edge of the lawn, dodging golfers and their balls. From August through April, watch for Kōlea. Grassy areas such as golf courses are favorite haunts for this bird. The clubhouse restaurant is open to the public, and is one of the few eateries in the area.

Mauna Loa Road

From Kīpuka Puaulu you can continue up Mauna Loa Road another ten miles to the road's end, at an elevation of 6,662 feet. There is parking, a picnic pavilion, and a pit toilet at the end of the road, and the trailhead for the grueling path to the top of Mauna Loa.

Mauna Loa Road is a good area for roadside birding, but there are only a handful of places to pull off the pavement over the length of the drive—so be sure to take advantage of them. One of the first places you can pull off is just a mile past Kīpuka Puaulu, at Kīpuka Kī. Under towering koa trees you should be able to spot the Red-billed Leiothrix, which is very common in this *kīpuka*. These birds chatter noisily at the sight of people, but will often allow you to approach quite closely.

All along the road, especially early in the morning or late in the day, watch for Kalij Pheasants. These birds have increased their numbers dra-

Kīpuka Kī

matically in recent years. It is not uncommon to spot ten of them on the way up the road. They will usually run through the shrubs to escape danger, but can fly to branches for security or night roosting. Other birds to watch for on the roadway or in roadside grasses are Nutmeg Mannikins, House Finches, and Eurasian Skylarks. Introduced birds you may see in the trees include Northern Cardinals and Japanese White-eyes.

Look for 'I'iwi in trees along the road. This bird has retreated from its former range around the Kīlauea Caldera, and Park Service biologists are not sure they know exactly why. Avian diseases such as malaria may play a role. Even up on the mountain this bird is not abundant. If you do not see 'I'iwi on the drive up or at the trailhead, you might stroll a ways up the Mauna Loa trail and hope that this bird's path crosses yours. 'Apapane, 'Elepaio, and Hawai'i 'Amakihi are other endemic birds that may be seen in the Mauna Loa Road area. Finally, watch for Pueo, the endemic Hawaiian subspecies of the Short-eared Owl, and 'Io, the Hawaiian Hawk. These birds prefer open areas where they can watch for prey from the air.

The Mauna Loa Road is sometimes closed due to high fire danger on this dry grassy mountain slope. Although there is very little traffic, the ten-mile drive is time consuming even if you don't stop for birds: the paved road is a good one, but very narrow and twisting.

Thurston Lava Tube

Molten lava tends to cool first on the surface, while the material below may continue to flow. Sometimes a stream of lava will flow right out from under the cooled crust above it, leaving a cave or lava tube. The most famous example of this in Hawai'i is Thurston Lava Tube, surrounded by lush 'ōhi'a forest in Hawai'i Volcanoes National Park.

Lava tubes, if undisturbed, provide a unique habitat for a variety of Hawaiian animals that have evolved to dwell within them, including crickets, earwigs, moths, centipedes and millipedes, and spiders. Many of these species are blind, the power of sight being useless in the inky depths of the lava tubes. Some cave dwellers are pale or white, since protective coloration is unnecessary against blind predators in the dark.

Millions of visitors have trekked through Thurston Lava Tube, and it no longer harbors the variety of subterranean creatures found elsewhere. It does feature easy access and lush surroundings, including birds, and it provides an introduction to this most unusual Hawaiian habitat. This area's popularity means that it will be on every tourist's itinerary and it will be crowded unless you come very early or late in the day. When two or three mammoth tour buses roll up and disgorge fifty people apiece, most of the birds will be temporarily frightened away.

To avoid most of the crowds, all you have to do is cross the road, where good birding trails lead down to the edge of Kīlauea Iki Crater, and along the crater wall to Kīlauea Iki overlook which is half a mile away. Along the trails on either side of the road, watch for 'Elepaio and Japanese White-eye in the understory. 'Ōma'o may be singing from well up in the trees, and 'Apapane may be working the lehua blossoms in the 'ōhi'a canopy or flying overhead. Most of the time Hawai'i 'Amakihi can be seen. The interpretive sign at the lava tube trailhead depicts an 'I'iwi, but these birds have been quite rare in the area for many years. The trail to Kīlauea Iki Crater offers views out over the crater. Anyplace in the park where you have such views, watch for White-tailed Tropicbirds.

The loop trail at Thurston Lava Tube is one of the few places where there is wheelchair access into the forest, if only for a short way. At the beginning of the trail is an overlook of the lava tube that allows good views of the treetops below. Canopy-dwelling 'Apapane and 'Ōma'o may be seen from this spot. Birders using wheelchairs can go left on the loop trail, opposite from the recommended direction of travel. Where the trail forks left again to the rest room, keep right. The first 50 yards of this paved trail are steep and a little rough, but after that you will have 250 yards of level, paved trail through 'ōhi'a-fern forest before reaching the steps that lead up out of the lava tube. Watch for birds of the understory such as 'Elepaio and Japanese White-eye.

Kīpuka Nēnē

A very dependable place to see Nēnē (in a more natural setting than the Volcano Golf Course) is Kīpuka Nēnē. As a bonus, the area has a campground and many hiking trails. To reach Kīpuka Nēnē, follow the Chain of Craters Road south from the Crater Rim Road 2.2 miles to Hilina Pali Road. At the well-marked junction, take Hilina Pali Road about five miles to the Kīpuka Nēnē campground and picnic area.

Most of the Nēnē here have been banded. If you look closely you may see a small radio transmitter mounted on a bird's back, with a thin gray antenna wire attached. Any encounter with Nēnē in this area is likely to provide you with a close look—the birds supplement their diet with campground handouts. They are so persistent that it is hard to refuse them, but conditioning the birds to accept food from people reduces their chances to successfully reestablish themselves in the wild.

Other species to watch for in this dry woodland include Common Myna, Northern Cardinal, Japanese White-eye, Erckel's Francolin, and perhaps an occasional Hawai'i 'Amakihi. Hawai'i's least-seen francolin species also lives along Chain of Craters Road. The Red-billed Francolin is distinguished by its bright red bill and legs and its yellow eye patch.

The area around Kīpuka Nēnē is closed while the geese are nesting, usually from November to March. You can still drive along Hilina Pali Road during this period.

Hōlei Sea Arch

Perhaps surprisingly, a national park famous for active volcanoes and 'ōhi'a-fern forests is also a good place to see marine birds. The Black Noddy is a resident of the sea cliffs along the Chain of Craters Road. This tern species, also called White-capped Noddy or Hawaiian Noddy, breeds on islands throughout the tropical Atlantic and Pacific. It nests on nearly all of the Northwestern Hawaiian Islands and several of the offshore islets, but there are few places on the main islands where it can be viewed at close range.

A similar species also found in Hawai'i is the Brown Noddy, but it is not known to nest on the main islands. The easiest way to confirm that you are looking at a Black Noddy is by the yellowish orange color of the legs and feet during the breeding season. Breeding is variable but is most common in spring. At other times the Black Noddy can be told from the Brown Noddy by the Black Noddy's gray tail and more pronounced white cap, but these can be difficult to discern.

A good place to see Black Noddies is at Hōlei Sea Arch, marked by a sign and turnout at the 18.9-mile point along Chain of Craters Road. An unmarked turnout at the 18.6-mile point is also a good spot to look for these birds. Both of these spots have stone walls along the cliff, providing a sense of security atop the fractured lava flows that drop abruptly into the sea.

The noddies nest in crevices in the lava cliffs. To find the nests, look for areas on the cliffs that have been stained by the birds' white droppings. The noddy gets its name from a characteristic nodding behavior sometimes seen in nesting colonies. The birds here scarcely have room to nod as they squeeze into crevices on the cliff face.

If you visit the area during the winter months, keep an eye out for the Bristle-thighed Curlew in grassy areas along the cliffs, and report any sightings to HVNP staff. This very uncommon winter visitor breeds in Alaska and winters on islands of the Pacific, including the Northwestern Hawaiian Islands. It has been spotted along this coast, and a few of them may spend the winter on the main Hawaiian Islands. A more common winter visitor here is the Kōlea. With a great deal of luck you might see a Laysan Albatross over the water.

Be sure to gaze seaward during your noddy-watching to look for dolphins and sea turtles that frequent these coastal waters. The high cliffs offer a good vantage point for spotting them. Dolphins form groups called pods that may contain dozens of individuals, but they can still be hard to see if they are swimming several hundred yards offshore. Their backs are often exposed as they swim at the surface, but they blend in with choppy water. If you are lucky, Hawaiian spinner dolphins may exhibit the behavior that gives them their name: leaping from the water, spinning on their tails, and crashing back into the ocean. It is quite a sight!

Although the cliff-dwelling noddies are very close to Chain of Craters Road, they cannot be seen from the roadway, and the cliff overlooks are not accessible by wheelchair. More accessible noddy viewing is available at Kehena, described at the end of this chapter in "Other Birding Spots."

It is worth noting that there is no longer a direct route up the coast to Kehena, only thirteen miles from Hōlei Sea Arch as the noddy flies. Since 1983, persistent lava flows have severed the road and destroyed a national park visitor center, as well as most of the nearby town of Kalapana. Even fairly new maps may be inaccurate as continuing eruptions cover roads and alter the landscape.

Nearby is another place that is well worth a visit. It has nothing to do with birds, but nevertheless I can't pass up the opportunity to recommend it. At the 16.4-mile point along Chain of Craters Road a sign marks the trailhead for the Pu'u Loa Petroglyphs. A level but rocky trail marked by *ahu* (cairns or stacks of rocks) leads you one-half mile to an area where hundreds of figures and designs have been carved into the lava. A wood boardwalk surrounding the site gives you a good viewing platform and protects the petroglyphs from your footfalls as well. One of the commonest designs is a hole in the lava surrounded by concentric circles etched in the rock. It is said that Hawaiian fathers placed the umbilical cords of their newborn into these holes to ensure a long life for their children. The light is best for viewing the petroglyphs in the morning or late afternoon. The hike can be a hot one at midday; carrying water would be a good idea.

'Ōla'a Tract

One piece of Hawai'i Volcanoes National Park lies well away from the other points of interest in the park, separated by Highway 11 and the town of Volcano. It is not discussed in the fine brochure and map that visitors receive when they enter the park. Rangers do not lead hikes in the area. This section of the park is the 'Ōla'a Tract, and it was acquired in order to protect rare endemic birds that once dwelled in it, including Hawai'i Creeper, 'Ākepa, 'Akiapōlā'au, and 'Ō'ū.

Sadly, none of these birds has been seen in 'Ōla'a in recent years. One of them, the 'Ō'ū, has not been seen anywhere on the Big Island since 1986. The range of the others has receded to private land and grounds of the Kūlani Correctional Facility to the northwest of 'Ōla'a. The reasons for this decline are not well understood, although avian diseases and natural die-back of 'ōhi'a trees at 'Ōla'a may play a part.

Between the 'Ōla'a Tract and the closed lands to the northwest, where rare endemics are more likely to be seen, lies a strip of land that is open to birders. This is the Pu'u Maka'ala Natural Area Reserve. This State reserve is blanketed with dense, wet 'ōhi'a-fern forest containing some of the tallest tree ferns to be found anywhere in the state, many

Ferns at Pu'u Maka'ala

towering over twenty feet. Despite the impressive plant life and the area's status as a natural area reserve, it has some drawbacks for birders. Although 'Ōma'o, 'Elepaio, 'I'iwi, and 'Apapane are all common in this forest, the chance of seeing any of the rarer endemics is very slim indeed. Getting into the area often involves some muddy hiking. And finally, the area is open to public hunting.

For intrepid birders who simply must visit Pu'u Maka'ala in the hope of glimpsing an 'Ō'ū, there are two ways to reach the reserve. From the Volcano area go north on Wright Road, Route 148, skirting the 'Ōla'a Tract and passing two ninety-degree jogs in the road, to the dead end 4.8 miles above Highway 11. A muddy trail leads up the hill through dense fern forest. The other approach is from the Stainback Highway, which branches from Highway 11 just south of Hilo. Follow Stainback to unmarked and unpaved Disappointment Road, which leads south at the 6.8-mile point. This point is near the upper end of an S curve in the otherwise fairly straight highway. Disappointment Road may be too rough and wet to drive except with 4WD, but you can walk it. From either approach, the birding at Pu'u Maka'ala is apt to be better on week-days when you are less likely to encounter hunters.

2. Saddle Road Birding Sites

FEATURES:
• Many endemic birds
• Fascinating botany, geology, and history

FACILITIES:

Some of the very best birding areas on the Big Island (or in the whole state, for that matter) are along the Saddle Road, Highway 200. The route passes through lush forest, bleak lava flows, and rolling grassland that provide diverse habitat for native and introduced birds. It is called the Saddle Road because it traverses the saddle or pass between Mauna Kea and Mauna Loa. Starting at sea level in Hilo, the road climbs gradually up to a summit elevation of about 6,700 feet and then descends the other side to join the Hawai'i Belt Road, Highway 190, near Waimea or Kamuela. Total length of the route is fifty-three miles.

This road has an undeserved reputation for being very rugged, brought on by car rental companies' refusal to allow their cars on the road. What's it really like? The road is fully paved from one end to the other. For nineteen miles from Hilo the route is quite good, more of a suburban street than a country highway. Near the nineteen-mile marker the road becomes narrower and the pavement rough and patchy for about four miles. Still, it is no worse than many country roads anywhere else. The remaining stretch of road down to Highway 190 is in fair condition. Sections of the road are widened and improved periodically, so eventually the whole length will be comparable in quality to other Big Island highways.

There are a couple of Big Island companies specializing in the rental of 4WD vehicles, and these firms allow you to drive the Saddle Road and unpaved backcountry roads. Look in the phone book, or check with the Hawai'i Visitors Bureau in Hilo for current information. Big Island residents drive passenger cars on the Saddle Road all the time and the cars do just fine. Most of the good birding sites along the way are right next

Housekeeping cabin at Mauna Kea

to the paved road or within hiking distance, so if you can make arrangements to use a 2WD vehicle legally, it may be all you'll need.

Birders who want to be close to the action along the Saddle Road can spend a few nights of solitude at Mauna Kea State Recreation Area, where the State has nice housekeeping cabins for rent at bargain rates. It's usually not difficult to get a reservation because few tourists know about the cabins and even fewer want to spend time in this high, dry country far from the ocean. There isn't much to do other than hike, observe the flora and fauna, and enjoy the isolation. To reserve a cabin, contact the Division of State Parks in Hilo or Honolulu (see appendix). The cabins are built on elevated foundations, so access for the disabled is challenging.

As mentioned in the chapter introduction, a permit from the State Division of Forestry and Wildlife is required to visit some Saddle Road birding spots.

When visiting the saddle, remember that you are at a very high elevation. When the sun is shining it can burn you quickly. Just as quickly, the sun can be obliterated by cold, windy, foggy rain. Even if it is sunny and warm when you leave your car for a hike, and you only plan to walk a mile or two from the pavement, take along rain gear. You will probably need it. Running back to the shelter of your car when a cloudburst catches you is not a good idea: the rough lava will skin you alive if you trip and fall. All the public facilities along the Saddle Road—drinking water, rest rooms, telephone—are at Mauna Kea State Recreation Area. Elsewhere, you're very much on your own.

With those warnings aside, here are some places where you can be flanked by two of the mightiest mountains in the world, surrounded by a geological and botanical wonderland, and see some of the rarest and most beautiful birds in the world. One can't ask for much more than that!

Tree Planting Road

At the 16.1-mile point on the Saddle Road (as measured from the Hilo end), watch for a small yellow forest reserve sign on the south side of the road marking the beginning of Tree Planting Road. This rugged 4WD road cuts straight through ten miles of 'ōhi'a forest, lava fields of various ages, and some pretty muddy places, coming out at the Stainback Highway. This area catches a lot of rain, so wet branches and tree fern fronds often overhang the rugged jeep trail. It is the first good area along Highway 200 to get away from the paved road and see several of the more common endemic species in a lush setting. Watch for 'I'iwi, 'Apapane, Hawai'i 'Amakihi, and 'Ōma'o. A permit is required.

Kaumana Trail

Another place to stop along the Saddle Road is at the 18.5-mile point, where the road makes an S curve. At the upper end of the curve is an obscure trail leading through the grass and 'ōhi'a trees south from the road. If you follow this for fifty feet you will intersect the old Kaumana Trail, at one time the access route from Hilo to the saddle area. The old trail is still well worn and marked by *ahu*. The State recreation map claims that you can pick up this trail at the 17.4-mile point on the Saddle Road and follow it to the 19.8-mile point. However, the two ends of the trail are hard to find, and the best birding is in the middle.

Follow the trail west (away from Hilo) as it gradually veers away from the road. Soon it will skirt a number of *kīpuka*. Watch for 'Apapane, 'I'iwi, Hawai'i 'Amakihi, 'Elepaio, and 'Ōma'o. Introduced birds include House Finch and Kalij Pheasant. Retrace your steps to return, making sure you do not pass the point where you entered the Kaumana Trail from the access trail.

Kīpuka 21

Some experts claim that this *kīpuka* offers the best roadside birding in all of Hawai'i, and it's hard to disagree. The *kīpuka* is a dense stand of koa and 'ōhi'a just to the north of the road at the 21.2-mile point, where the powerlines cross over the highway. It is lower than the road, which is built on a younger *'a'ā* flow that approached the old forest and then stopped. Thus, you can view the *kīpuka* from the treetop level, giving you a bird's-eye view of the many endemic forest birds that occupy the canopy.

The best viewing is from the roadside, but if you want to see what a relatively undisturbed *kīpuka* looks like from ground level, you can walk down the powerline access road that lies just beneath the wires until you get to the edge of the vegetation. It is difficult to get very far into the *kīpuka* because it is rugged and dense.

Fog, mist, and rain are fairly common in the saddle area, but if the weather cooperates you should have no trouble seeing Hawai'i 'Ama-kihi, 'Apapane, 'I'iwi, and 'Ōma'o from the roadside. Japanese White-eye and House Finch are also common. The 'Elepaio is another resident of the *kīpuka,* but this bird frequents the understory and is difficult to see from the road. If you walk down to the *kīpuka* you may spot one. It is conceivable that a Hawai'i Creeper, 'Ākepa, 'Akiapōlā'au, and even an 'Ō'ū could be seen, although you would have to be quite fortunate to spot these rarities.

Along the road, notice that a gray lichen covers much of the black *'a'ā* lava. Lichens are plants that are composites of algae and fungi. Both plant partners profit from this symbiosis: the fungus benefits from the photosynthesis of the alga, and the alga benefits from the ability of the fungal walls to catch and hold water. This lichen will often hold water like a sponge even when the surrounding rocks are dry. This means you usually can't sit on these rocks while you scan the treetops for birds or you'll go away with a wet seat. Please don't ask me how I know this.

Pu'u 'Ō'ō Trail

A wealth of birdlife, gentle terrain, fascinating geology and botany, and absolute solitude make the Pu'u 'Ō'ō Trail one of my favorite birding spots anywhere. The trail is an old route used to drive cattle down from high grazing land to the Volcano area. The route weaves its way south from the Saddle Road for nearly four miles across fairly level ground to a point where the trail merges with Powerline Road. It is one of the most interesting trails in the saddle area because it passes through several dis-

Along the Pu'u 'Ō'ō Trail

tinct geological and ecological areas. The Pu'u 'Ō'ō Trail is also a place where you will have an excellent chance of sighting many of Hawaii's endemic birds, perhaps even some of the very rare ones.

The trail is named for the nearby *pu'u* or hill, and the hill is named for the Hawai'i 'Ō'ō, a large (twelve-inch) honeyeater that was once common on the island. This glossy black bird had yellow undertail coverts (the feathers on the underside of the tail) and axillary plumes (tufts near the shoulder of the bird). This striking plumage made it a target for the feather trade, and it is reported that as late as 1898 more than a thousand of them were taken from the forests above Hilo. In addition to hunting, habitat modification almost certainly harmed the birds as well, and early in this century their numbers decreased drastically. There were unconfirmed sightings into the 1970s on windward Mauna Kea, but today most biologists believe the Hawai'i 'Ō'ō is extinct, gone forever from the hill and trail that bear its name.

The Pu'u 'Ō'ō trailhead is a little hard to find. It is on the south side of the road, marked by a sign at the 22.4-mile point, but the sign is about fifty yards off the road and you have to watch for it. The trailhead is 0.2 miles west of an old rusty roadside sign that says "Mauna Kea Soil Conservation District." Across the road from the trailhead is an old wooden gate through the cattle fence.

The route is fairly well worn and marked by a variety of yellow plastic cards, colored plastic ribbons, and *ahu*. Still, the trail can be a bit hard to follow where it passes through old, smooth, and sparsely vegetated *pāhoehoe* lava flows. Be sure to observe your route carefully so you can find the trail again on your way out.

The trail passes through the Kīpuka 'Āinahou State Nēnē Sanctuary, but the Nēnē have not prospered here. The reasons may include habitat modification, introduced predators, diseases, competition from other species, and even illegal hunting. In order to protect the birds during their breeding season, the sanctuary and the section of the trail passing through it are closed to hunters from November 1 through February 28, but birders and hikers may use the area year-round.

The first 150 yards of the trail pass through rough, crumbling '*a*'*ā* lava that is difficult to walk on. Next you will descend through 'ōhi'a scrub, where you have an excellent chance of seeing Hawai'i 'Amakihi, 'Apapane, 'I'iwi, and 'Ōma'o. At about the half-mile point the trail reaches grassland and open forest over old *pāhoehoe* lava. Listen for the chattering of the Red-billed Leiothrix and California Quail. You may also see Kalij Pheasants, Wild Turkeys, and Ring-necked Pheasants in the grass.

After another three-quarters of a mile you will reach an area where there are many blackberry plants in the grassland. When the berries are ripe, expect to see House Finches feeding on the fruit. At this point, there is an area of large koa and 'ōhi'a trees about a hundred yards to the right (west) of the trail. A little farther on, the trail passes directly through a thick stand of koa. Spend plenty of time at this spot. Most of the little green birds you see in this *kīpuka* will be Hawai'i 'Amakihi or Japanese White-eyes, but an extremely lucky birder might see an 'Akiapōlā'au, Hawai'i Creeper, or 'Ākepa (the adult male 'Ākepa is red-orange). Another endemic to watch for in the *kīpuka* is the 'Elepaio.

When you climb out of the koa thicket, you will be faced with the desolate '*a*'*ā* flow of 1855. Past this there is another mile to mile and a half of 'ōhi'a scrub and grassland before you reach a large, dense *kīpuka* with very good birding. Beyond here the trail gets rougher and more difficult to follow. Energetic birders can continue on to another *kīpuka* between three and four miles from the road where the rarest of the area's endemics are most likely to be seen. (This *kīpuka* is only about a hundred yards west of the Powerline Road described below.)

Older maps show the trail extending 6.5 miles south from the Saddle Road, to the point where the trail reaches private land. Today, the last third of the trail is obliterated by the lava flows of 1984. Toward the end of the trail it is *extremely* difficult to discern. Note your surroundings frequently to avoid becoming lost. Eventually the trail merges with Powerline Road if you can follow it that far. You can retrace your steps or return to the highway on Powerline Road for a change of scenery. From the Powerline trailhead it is about 0.8 miles to the Pu'u 'Ō'ō trailhead. Total loop distance is about nine miles. A permit is required.

Powerline Road

Near the Pu'u 'Ō'ō Trail and roughly parallel to it is Powerline Road, a 4WD trail forged along a powerline south of the Saddle Road. The trailhead is marked by a bullet-pocked sign reading "PLR" located at the 21.6-mile point on the Saddle Road. The powerline and poles have been removed, with only pole stumps dotting the landscape along the road for most of its length.

This very rugged road is less scenic than the Pu'u 'Ō'ō Trail, but it allows you to drive near *kīpuka* off the highway with a 4WD vehicle if you are a skilled off-road driver. The rough road is fairly level but somewhat difficult to walk on because of the loose crushed lava that covers much of it.

The first *kīpuka* is less than a mile from the trailhead, and thereafter the road passes through or near *kīpuka* every quarter mile or so for several miles. Be very careful if you walk over lava flows to *kīpuka* off the road. Much of the lava is loose, and small lava tubes can collapse under your weight.

The bird to watch for in this area is the 'Ō'ū. Powerline Road itself is probably at the upper elevation of 'Ō'ū range, but a 1979 survey of forest bird communities reported as many as twelve 'Ō'ū at a single observation station in a *kīpuka* below (to the east) of the road. Based on these observations, the 'Ō'ū population in the area was estimated to be four hundred birds. Since that time the situation has changed tragically. The last time anyone saw an 'Ō'ū on the Big Island was in 1986. If you have the rare good fortune to glimpse one of these critically endangered birds, be sure to report it to State or federal authorities.

Birds you are likely to see include Hawai'i 'Amakihi, 'Apapane, 'I'iwi, and 'Ōma'o. In the *kīpuka* watch for 'Elepaio. Rare birds to watch for include Hawai'i Creeper, 'Ākepa, 'Akiapōlā'au, and Nēnē. Introduced species include Japanese White-eye, House Finch, Red-billed Leiothrix, California Quail, Kalij Pheasant, Wild Turkey, and Ring-necked Pheasant.

Inside a *kīpuka*

Hakalau Forest National Wildlife Refuge

On the northeast flank of Mauna Kea, above the zone where mosquitoes imperil native birds and below the rolling grasslands where cattle have long since removed most other vegetation, there is an expanse of wet forest where the rarest Big Island birds still survive. This rugged land is nearly inaccessible to livestock and to people, and the resulting lack of habitat modification has left endemic forest birds with the relatively intact ecosystem they need to survive.

The area is blanketed by a huge expanse of wet forest dominated by koa, ʻōhiʻa, and tree ferns. In this forest are the highest population densities of ʻĀkepa and Hawaiʻi Creeper to be found anywhere, and very large populations of many other Hawaiian forest birds such as ʻIʻiwi, ʻApapane, ʻElepaio, ʻŌmaʻo, and Hawaiʻi ʻAmakihi. Smaller numbers of ʻAkiapōlāʻau are also present. To make the area even more tantalizing, there is history and fascinating ecology here, and splendid views when the weather is clear.

Much of this area is protected in a patchwork of federal and State land parcels, the most important being the Hakalau Forest National Wildlife Refuge. This is the only national wildlife refuge in the country established for forest birds. The small staff at Hakalau battles to minimize the effects of plants and animals that have invaded this forest, erecting fences so no more stray cattle and feral pigs can enter the refuge, and destroying introduced plants that have already invaded.

The presence of rare and wonderful birds is the good news. Unfortunately, there are also some major drawbacks for birders. First, you need to time your visit carefully. Only one part of the refuge is open to birders, and it is open only during the last weekend of the month. Access to the public is limited because the staff of the refuge is just too small, and stretched too thin, to protect the birds and cater to visitors. There are no services or facilities for visitors at the refuge.

A visit to Hakalau also requires some dedication. The refuge can be reached only by traveling some sixteen miles on a rugged 4WD trail off the Saddle Road, including dusty stretches on the south side of Mauna Kea and muddy areas closer to the refuge. The drive from Hilo will take a couple of hours, and at the end of the trip your birding zeal may be diminished. Once you get to the area, expect weather that is often cold, rainy, and just generally foul.

If you plan to visit, be sure to make reservations. They may be obtained by calling the refuge office in Hilo between 8:00 A.M. and 4:00 P.M., but you must call three Wednesdays before the day you plan to visit (see appendix). A brochure is available that describes the rules you must follow in the refuge.

If a solo visit sounds a bit intimidating, there's another way you might be able to arrange a trip to Hakalau. Volunteer work parties occasionally visit the refuge for a long weekend. After controlling alien plant pests, mending fences, collecting koa seeds, planting trees, or perform-

Collecting koa seeds at Hakalau

ing other hard work at the refuge for a day or two, there is usually some time for birding with refuge biologists. These work groups are composed largely of Hawai'i residents and are set up by organizations such as the Hawai'i Audubon Society or the Hawai'i Chapter of the Sierra Club. If you are planning to visit Hawai'i, contact these groups well in advance of the season when you plan to visit. If your schedule is flexible and you don't need to be pampered, you may be able to visit the refuge and help maintain it as well.

To reach the Hakalau area, take the Saddle Road to Mauna Kea Road, at the twenty-eight-mile marker on Highway 200. This road is easy to spot—it's an excellent paved road, and it starts just opposite a small cabin that serves as a State hunter checking station. Just 2.1 miles north of the Saddle Road, unpaved Keanakolu Road branches to the right toward Hakalau. Watch for a cattle guard right at the beginning. If you drive this route early in the morning you will know you are heading east, because you will be driving right into the sun. This road leaves paved Mauna Kea Road just before the paved road reaches a dense stand of pine trees and a large sign with information about Mauna Kea. Set the trip odometer on your vehicle when you leave the pavement so you can measure your progress. Keep in mind that the distances in this description should be considered approximate because every odometer reads a little differently, especially over the long distance to Hakalau. The dirt road is fairly well maintained, although you will need high-clearance 4WD to pass a few rocky gullies. The road skirts the flank of Mauna Kea through open cattle range all the way around to the Hakalau area on the east-northeast side.

At the beginning the landscape is nearly barren except for grass and clumps of gorse, a spiny introduced shrub that the grazing cattle won't touch. Gorse seeds first hitchhiked to Hawai'i in the wool of sheep brought from Europe. As you bounce along on the trail you will begin to see a few big, lonely old koa trees. All the young unprotected koa seedlings are mowed down by grazing cattle, so these patriarchs are likely to be the last koa that will ever grow in this area without some human help, such as fence exclosures, to keep cattle away from seedlings. Some of the roughest parts of the road are the most historically interesting. You will pass some sections where cobblestones are visible in the road. These are remnants of the old Hilo-Waimea road. It was built up on the mountain rather than down along the coast near the location of the modern highway to avoid the task of building bridges across the many gulches of the Hāmākua coast.

There are some birds to be seen in this open country. Koloa are fairly common inhabitants of the stock ponds you pass along the road. Eurasian Skylarks are also common along the road. Wild Turkeys and Kalij Pheasants may be seen, especially in the more forested areas, and Pueo and 'Io, the Hawaiian owl and hawk, are sometimes seen soaring overhead. Other common birds include Erckel's Francolin, California Quail (most visible early and late in the day), and Kōlea. The Ruddy Turnstone is a less common visitor. Two quail species that occur in these pastures but are not likely to be seen are Gambel's Quail and Japanese Quail. Gambel's is quite rare, and the tiny Japanese Quail is nearly impossible to flush from cover.

As you drive around the mountain the koa become increasingly dense until they outnumber the cattle, and the terrain is more forest than grassland. From here you will pass several parcels of State and federal land. A little more than fifteen miles in you will pass a narrow strip of State land, the Pīhā Section of the Hilo Forest Reserve, managed by the State for hunting. Watch for a yellow sign on the fence on your right. With a permit you can enter this hunting area to look for birds, but you would probably see more endemic birds at many places along the Saddle Road that are far more accessible. This State land is overgrown with introduced banana poka and harbors very few of the uncommon endemic birds. If you enter this area, be careful not to lose your way on the network of hunting trails.

At this point you have already passed part of the National Wildlife Refuge, but the boundary is some distance from the road and you will not have seen the signs. At about 16.5 miles you will come to the gate of the Maulua Tract of the refuge, marked with a prominent sign. This is the area open to birders and hikers. You can pass through the gate, making sure to close it behind you, and drive down rough Maulua Road for about three miles to a gate at the bottom of the road. From this point the forest becomes increasingly dense as you follow trails downslope. Be sure to observe your surroundings so you can find your way back, and remember that the hike back to your vehicle will be mostly uphill. You are a long way from help, so make sure you are prepared to cope with bad weather, darkness, or injury.

If you continue driving on Keanakolu Road you will reach the Laupāhoehoe Section of the Hilo Forest Reserve at about 16.9 miles. Finally, at 17.7 miles, you will see one of the familiar Hawaiian Warrior signs of the Hawaii Visitors Bureau. You're not hallucinating from trail dust. In 1834, the famed naturalist David Douglas died near this spot under suspicious circumstances. (His body was found in a pit meant to trap feral cattle. He may have fallen in accidentally, but some believe he was murdered by a cattle hunter who robbed him.) There is a short trail from the HVB sign down to a monument. A stand of Douglas fir, a tree species named for the naturalist, has been planted nearby. There are also introduced eucalyptus and redwood near the site. You can continue on this road as it circles Mauna Kea, but you may eventually be stopped by locked gates.

Puʻu Huluhulu

Right at the junction of the Saddle Road and Mauna Kea Road is Puʻu Huluhulu, a hill that became a *kīpuka* when lava flows surrounded it. This *kīpuka* harbors some native birds, is rich in endemic plant life, and has a 0.5-mile trail leading up to a nice picnic spot at its summit.

The trail starts on the east side of the hill, where a red lava road heads south from the Saddle Road. There is space to park, and an opening in the fence that surrounds the hill. Along the hike up the hill, watch for Hawaiʻi ʻAmakihi, ʻApapane, and ʻIʻiwi. Among the endemic plants that survive in the area are sandalwood and ʻākala, the native raspberry. Unfortunately, an introduced plant called German ivy is common in the *kīpuka*. Its bright green, glossy leaves are impossible to miss along the trail.

If the trail has not been maintained recently, it may be a little hard to follow in the dense grass. Be sure to watch your step on the steep hill, especially where grass may conceal irregularities in the trail.

Puʻu Lāʻau

To ornithologists in Hawaiʻi, the name Puʻu Lāʻau is almost synonymous with the Palila, an endangered finchlike honeycreeper found only on the slopes of Mauna Kea. Although the range of this bird extends in a rough belt around half the circumference of the great mountain, Puʻu Lāʻau boasts by far the highest population density. This area, above seven thousand feet on the western or leeward slope of Mauna Kea, is grassy woodland dominated by the māmane, a native tree of the legume family. (Puʻu Lāʻau means forested hill.)

The māmane is vital to the survival of the Palila. The birds' principal food is green māmane seeds, and the Palila population appears to fluctuate in the range of 1,600 to 6,400 individuals according to the abundance of māmane seeds. Occasionally the birds also take insects or berries from the native naio tree, also common in the Puʻu Lāʻau area.

The turnoff for Puʻu Lāʻau

Until recently this area was also home to large numbers of mouflon sheep and feral ungulates—sheep and goats whose domestic ancestors were released and proliferated. These voracious creatures grazed on all the māmane branches they could reach and nibbled off most new seedlings, putting the food supply of the endangered Palila in peril. Environmental groups went to court to force the State to remove the animals from the area so that the Palila would not be starved to extinction. In response, the State has allowed nearly unrestricted hunting in the area so that the population of sheep, goats, and mouflon are not sustained. This court action has caused hard feelings between some birders and some hunters, because the once popular hunting area now has very low levels of game animals and relatively poor hunting. Keep the dispute and the hard feelings in mind when birding here.

To reach the area, take the Saddle Road to the 43.3-mile point where the unpaved access road heads north toward Pu'u Lā'au. There used to be a hunter checking station just across the highway from the beginning of the road, and it has been described or depicted as a good landmark in some guidebooks. This checking station has been torn down and replaced with a new concrete block structure a few hundred yards off the highway on the dirt road to Pu'u Lā'au. Fortunately, there is a prominent sign on the highway that points the way toward the "Kilohana Hunter Checking Station."

Recently the road to Pu'u Lā'au has been graded and graveled, making it dusty but passable even in a passenger car unless rain has washed gullies into the roadway. The hill called Pu'u Lā'au is 4.1 miles north of the Saddle Road along this access road, near a cabin and a grove of eucalyptus trees. This point is also the boundary between two State hunting units, A and G. You will see signs to this effect near the cabin. This is a good place to park and start walking. Just before you reach the cabin, at 4.0 miles, is a game-watering station. This is a good place to check for birds.

The best place to look for Palila is about a mile past the cabin. Follow the road that passes through the eucalyptus grove behind the cabin, to a point where a side road bears to the right. Stroll some of the many sheep trails that cut through the grass. With any luck, you will see Palila in the māmane looking for green seed pods. Another uncommon endemic that might be seen is 'Akiapōlā'au. Other endemics to watch for include Hawai'i 'Amakihi throughout the area, 'Elepaio in thick stands of trees mostly below the cabin, and 'Apapane and 'I'iwi when the māmane trees are in bloom, especially during September and October. Introduced bird species include House Finch, California Quail, Ring-necked Pheasant, Erckel's Francolin, Eurasian Skylark, and Wild Turkey.

The introduced grass at Puʻu Lāʻau that is no longer eaten by sheep and goats creates a serious fire hazard when it is dry. Be very careful with fire, and do not park a hot vehicle right over tinder-dry vegetation. A permit is required to visit Puʻu Lāʻau.

3. Manukā Natural Area Reserve

FEATURES:
• Abundant endemic birds

FACILITIES:

The drive between Volcano and Kailua-Kona along the south side of the island is ninety-six miles, a little long for a nonstop stretch at the wheel. About midway between the towns, at the 81.1-mile point on Highway 11, is a pleasant little park and rest stop where travelers can take a break. Manukā State Wayside and Natural Area Reserve offers rest rooms, picnic tables, pavilions, and nonpotable water. There is also a large lawn area that ends abruptly at the surrounding forest of tall ʻōhiʻa trees and introduced understory plants that include guava. The air is a little cooler at this eighteen hundred-foot elevation than it is at sea level along the sunny Kona coast. Altogether, it is quite a pleasant little spot. The only minor drawback is the presence of some hungry mosquitoes, but they are easily dispelled with your trusty repellent.

Conventional wisdom would say that this is no place to find endemic forest birds. The elevation is too low, the mosquito count is too high, and there isn't another accessible place to see endemic birds for fifty miles in either direction. Nonetheless, they are lurking just inside the forest that surrounds the lawn areas of this park. Hawaiʻi ʻAmakihi even venture out of the forest to feed on the ornamental flowers near the parking lot.

Fortunately for birders, the State land at Manukā is actually far more extensive than the developed area around the wayside, stretching from the sea up to the five thousand-foot level in the mountains of Kaʻū. There is access to the forest via a two-mile loop trail that ascends gradually from the wayside up into the dense stand of 'ōhi'a, providing the perfect opportunity to view the birds. You may spot 'Apapane, Hawai'i 'Amakihi, and 'Elepaio before you have hiked a hundred yards. Introduced species to watch for include Japanese White-eye, Red-billed Leiothrix, Northern Cardinal, Common Myna, and Kalij Pheasant.

Recently the State of Hawai'i prepared an excellent brochure and trail guide on Manukā. Pick one up from the box at the trailhead, or get one in advance from the Division of Forestry and Wildlife Baseyard in Hilo (see appendix). The loop trail starts in the 'ōhi'a trees just uphill from the parking area. At the very beginning of the trail a dirt road branches to the right and leads to several water tanks; the loop trail you want continues straight up the hill at a fairly gradual climb. The path is generally very easy to follow, but it can get somewhat overgrown if the State has not cleared it for some time. Numbered markers along the trail identify points of interest that are described in the trail guide. Good shoes are recommended because some parts of the route are laced with protruding roots or loose, crushed lava.

About one mile up the trail you will pass a large pit on your right. This was formed by the collapse of a lava tube. It is a beautifully lush spot, so stop to have a look. Be careful not to get too close to the pit, as it would be very difficult to get out—if you survived the fall. At this point the trail veers left for several hundred yards, and then starts back down the slope. At the bottom of the hill, you will come out of the forest a few hundred yards west of where you started. Cross the lawn back to the parking area.

Why are there endemic forest birds in this seemingly unlikely spot? The answer may lie in the towering presence of Mauna Loa, which translates as "long mountain." Mauna Loa isn't quite as high as Mauna Kea to the north, but it makes up for it in bulk. The mountainsides rise sharply up to elevations that are above the range of disease-carrying mosquitoes, and this high-elevation avian refuge extends for forty miles to the other side of Hawai'i Volcanoes National Park. There are relatively few roads into this area, so disturbance is minimized. The birds at Manukā probably spill over from the extensive forests upslope. These birds may represent populations that are developing immunity to the diseases carried by mosquitoes, or hapless migrants whose wanderlust places them at risk in the mosquito zone.

Access for the disabled is limited at Manukā. The rest rooms are wheelchair accessible, but the rough trail is not. There is a sprawling subdivision next to Manukā (Hawaiian Ocean View Estates) with roads up into the forest, but these roads are private and they do not lead into the reserve. Birders may not be welcome in this area.

Another interesting feature of Manukā Natural Area Reserve is the shape of the land parcel itself. It is shaped like a piece of pie, wide at the coast and narrowing to a point in the mountains. This is a traditional land division of Hawai'i, called an *ahupua'a.* The reserve is named for the Manukā *ahupua'a,* which it occupies. These self-sufficient divisions typically encompassed several habitat types and vegetation zones, providing for all the needs of their human inhabitants. From fish at the shore to giant koa logs in the interior, the residents of an *ahupua'a* found virtually all they needed to carry on their way of life.

4. Whittington Beach County Park

FEATURES:
• Marine and wetland birds
• Sea turtles

FACILITIES:

Another worthwhile stop along Highway 11 between Volcano and Kailua-Kona is Whittington Beach County Park, where a large pond offers the possibility of spotting wetland birds. The park is at the 60.5-mile point on Highway 11, near the town of Na'alehu. Watch carefully for the entrance, because the sign is small and easy to miss. To reach the pond area, keep to the left about 0.2 miles in from the park entrance, and follow any of several dirt roads to the water. You might see Kōlea, Wandering Tattlers, Black-crowned Night-Herons, ducks, or other surprises here, especially in winter. The area is heavily used by local anglers, so most birds are frightened off before they have stayed for long.

If you follow the entrance road all the way into the park you will reach the parking lot, rest rooms, and picnic pavilions. Look for White-tailed Tropicbirds overhead in this area. Introduced species to watch for in the trees and around the picnic area include Japanese White-eye, Zebra Dove, Spotted Dove, Northern Cardinal, and Common Myna. Also look for mongooses as they raid the picnic area trash cans.

Green sea turtle

The collapsed piers at Whittington are remnants of the days when Honuʻapo Bay was a sugar port. The surf is always treacherous and the shore is rocky, but despite the dangers there are little mariners who still use the bay: watch for sea turtles as they bob around just offshore, feeding on seaweed. They can be spotted as they raise their heads out of water to breathe, or when their shells protrude from the water surface. Most common is the Hawaiian green sea turtle, but occasionally a Hawaiian hawksbill turtle may be seen.

A better place for turtle watching is at Punaluʻu Beach County Park, five miles up the road toward Volcano. This spot, famous for its black sand beach, has calmer waters and offers vantage points well above sea level to facilitate your viewing. I have never failed to spot turtles at Punaluʻu. Try the rocky area near the rest rooms at the southwest end of the beach, or the old pier ruins at the northeast end. Most of the turtles in Hawaiian waters nest in the Northwestern Hawaiian Islands, where they are protected within the Hawaiian Islands National Wildlife Refuge. However, a few females haul themselves up on the sands at Punaluʻu to dig their nests and lay their eggs. Look for nest sites that have been fenced off to protect them from predators and careless humans.

5. Loko Waka Pond

FEATURES:
• Wetland birds

FACILITIES:

Hilo residents have an important wetland right in their backyards—literally. Loko Waka Pond, privately owned and used by local residents for rearing edible mullet and milkfish, is right at the edge of town. The pond is spring fed and connected to the sea by underground channels. Although the pond is private, there is good viewing from the shoulder of Kalaniana'ole Avenue. This is a dependable place to see Hawaiian Coots and Black-crowned Night-Herons. Cattle Egrets can also be seen.

To get to the pond, go east on Kalaniana'ole Avenue (Highway 137)

Loko Waka Pond

for 2.5 miles from the junction of Highways 19, 11, and 137. This junction is at the eastern end of Banyan Drive, the major hotel district in town. Park along the *mauka* shoulder of the road at the edge of the pond, or across the road at James Kealoha County Beach Park. Exercise caution as you view the birds. The shoulder of the road is narrow and traffic moves quickly.

Two endangered wetland birds that you might expect to see at the pond are absent. The Common Moorhen has not been found on the Big Island since the early 1900s. The Black-necked Stilt can be seen at 'Aimakapā Pond on the west side of the island, but has been recorded at Loko Waka Pond only once.

A few winter migrants visit the area. Northern Pintails and Northern Shovelers might be seen, although their numbers are declining in Hawai'i and they are generally more common at 'Aimakapā Pond. Less common migrants to watch for include American Wigeon, Lesser Scaup, and Ring-necked Duck.

Right in the center of Hilo is Waiākea Pond, an estuarine pond at the mouth of the Wailoa River. The pond is the centerpiece of Wailoa State Park, an urban park where lawns surround the water. Many local people fish here for mullet, milkfish, āholehole, and pāpio. Most of the birds in the area are introduced urban birds. Expect to see Common Mynas, Japanese White-eyes, Northern Cardinals, House Sparrows, flocks of up to a hundred Nutmeg Mannikins on the lawns, and a variety of domestic ducks and geese, including Mallards. In winter some interesting ducks and geese may stop at Waiākea, including the species seen at Loko Waka Pond.

6. 'Aimakapā Pond

FEATURES:
• Wetland birds

FACILITIES:

One of the most important wetlands in Hawai'i is near Kailua-Kona at 'Aimakapā Pond, just north of the Honokōhau Harbor. It is a dependable

'Aimakapā Pond

place to see the Hawaiian Coot and Hawai'i's endemic subspecies of Black-necked Stilt. A variety of other wetland migrant species turn up as well. To get to the pond, drive north from Kailua-Kona about three miles on Highway 19 to the Honokōhau Harbor. The entrance road to the harbor area is well marked. Keep to the right on the entrance road, passing the restaurant and concession area. This road curves around the harbor past many old weathered boats propped up on stilts. At the end of the pavement there is a small parking area of rough gravel.

The area around 'Aimakapā Pond attained an important level of protection when the area was set aside as the Kaloko-Honokōhau National Historic Park. An unmarked trail heads north to the pond from the parking area. Follow it across a grassy field in a direction roughly parallel to the shoreline and the highway. Do not turn left at the branch that is seventy-five yards from the parking lot; this leads to a private residence. Continue several hundred yards from the parking lot, turn left at a fairly prominent path, and head through a thicket of large kiawe trees that surrounds a new rest room building. Finally, turn and head up the rocky beach to the sandy area that separates 'Aimakapā Pond from the ocean, a total hike of about one-third of a mile.

The trail is well worn, but not by armies of excited birders looking for wetland birds; the narrow sandy area between the pond and the ocean is the Kona area's unofficial nude beach. The best time for birders to visit is two or three hours after sunrise, before the area is too crowded. If you come earlier you will be looking into the sun, and if you arrive later you may be distracted by the moon!

At the beginning of the access trail watch for Nutmeg Mannikins. While walking through the kiawe, species to look for include Northern Cardinal and Yellow-billed Cardinal, Zebra Dove and Spotted Dove, House Finch, Rock Dove, and Japanese White-eye. The Lavender Waxbill appears to be increasing in this area. The Black Francolin is another uncommon resident. Along the rocky coast you may see Ruddy Turnstone, Kōlea, and Wandering Tattler. These birds are normally migratory, but a few of them can be seen all year. Another common winter shorebird to watch for is the Sanderling.

At the pond, year-round residents include Black-necked Stilt, Hawaiian Coot, and Black-crowned Night-Heron. Cattle Egrets are uncommon but quite visible when present. There are no recent records of Common Moorhens on the Big Island. A fairly new resident of 'Aimakapā Pond is the Pied-billed Grebe. The dozen or so you may see are the descendants of a pair that arrived in 1985. Artificial floating platforms placed in the pond are used as nest sites by Black-necked Stilts and Hawaiian Coots.

In the winter watch for an array of migrant waterfowl. Numbers of migrants are variable and have been declining, but you may see Northern Pintail and Northern Shoveler. Any of the uncommon waterbird stragglers that find their way to Hawai'i, including ducks, gulls, and terns, may turn up at 'Aimakapā.

Kaloko Pond, less than a mile to the north, sometimes harbors shorebirds, but ducks are seldom seen there.

7. Kaloko Drive

FEATURES:
• Endemic forest birds near Kailua-Kona

FACILITIES:

Most visitors to Kailua-Kona don't realize they are just ten miles from the lofty peak of 8,271-foot Hualālai. A mountain this large seems hard to overlook, but the ocean tends to draw one's gaze away from the land, and volcanic haze from Kīlauea often obscures views along this coast. Mauna Kea to the northeast and Mauna Loa to the southeast are so much larger and more accessible that they dwarf Hualālai and grab most of the publicity.

Still, like other great mountains in Hawai'i, Hualālai provides refuge for a few endemic bird species that have managed to persist on the high slopes. The high elevation affords some protection against mosquito-borne diseases, and for a long time the terrain provided protection from human interference as well. Unfortunately, the steep subdivision road that now provides birders with access to the mountain has also brought threats to already declining bird populations in the form of alien plants, pets, and increased human activity. Despite this, the adaptable Hawai'i 'Amakihi and the relatively abundant 'Apapane can be seen in the area. With luck and persistence you may even spot an 'I'iwi along this road.

To reach this area from Kailua-Kona, follow Palani Road up the hill from the center of town. It becomes Highway 190, the main road to Waimea. Just one mile past the junction with Highway 180, at the 34.2-mile point on Highway 190, is the intersection with Kaloko Drive. Although the road is marked by signs, it can be easy to miss because the traffic moves so quickly. Paved Kaloko Drive climbs steeply up the side of dormant Hualālai, which last erupted in 1801. There are many paved side roads that branch from it, but the main road is always easy to follow. The route climbs for over six miles, reaching an elevation of about five thousand feet on the slope of the mountain.

Few homes have been built in the subdivision yet and that is probably one reason why there are still a fair number of endemic birds in the area. In addition to Hawai'i 'Amakihi, 'Apapane, and 'I'iwi, watch for 'Io. No one has seen an 'Alalā on Hualālai since 1991. Introduced birds include Northern Cardinal, Japanese White-eye, and House Finch in the forest, and Yellow-fronted Canary, Saffron Finch, and Common Myna in more open areas. Less common are Nutmeg Mannikin, Warbling Silverbill, and Red-billed Leiothrix. Also watch for Erckel's Francolin, Kalij Pheasant, and Common Peafowl.

As you drive this road, consider that it was slashed into fairly pristine native forest. There are plenty of other places to build homes in the area, but this forest is unique, dwindling, and irreplaceable. When the forest is destroyed or greatly disturbed, many species face no alternative but extinction. These Islands desperately need more voices to stand up for the land and the birds and the forests. Kaloko Drive is a grim symbol of our disregard for natural Hawai'i and the life that existed here long before humans ever reached these shores.

8. Pu'u Anahulu Area

FEATURES:
• Extremely wide variety of introduced birds

FACILITIES:

Wildlife managers in Hawai'i are very reluctant to allow new species to be introduced to the Islands. History has shown that introduced birds can compete with native species, harbor avian diseases, and become agricultural pests. This prohibition on bringing alien species to the Islands is fairly recent, however, and the previous willingness to introduce birds has resulted in the diversity of birdlife we see in Hawai'i today. One spot that is particularly rich in introduced birds is the Pu'u Anahulu area along Highway 190 between Kailua-Kona and Waimea. The avian diversity in this area is due to a program of authorized and unauthorized introductions made by the former owners of a large ranch in the area. For several decades the State cooperated with the ranch owners to introduce gamebirds and encourage their establishment through exclosures with feeding and watering stations where the birds would be protected. Apparently many cagebirds were introduced as well, and the result is quite an odd mix of birds.

Birders are hampered by a scarcity of roadside turnouts and public lands along this stretch of road. Several species are locally common but may be difficult to find because of poor access. Still, there are a few places worth a stop if you are driving through the area. Just south of the tiny town of Pu'u Anahulu, the highway makes a long S curve as it traverses a hill. At the 20.8- and 20.9-mile points there are turnouts where you can stop if you are headed southbound toward Kailua-Kona. This is a good spot to look for introduced Estrildid finches, including Lavender Waxbill and Red Avadavat. Red-cheeked Cordonbleu and Black-

Hunter checking box at Puʻu Anahulu

rumped Waxbill are other species that have been seen in the area but are quite rare. The House Finch is far more likely to be seen than any of these little cagebirds.

At the top of the hill is the private subdivision of Puʻu Lani, with the entrance at about the 20.4-mile point on the highway. This development used to be a good place to look for these Estrildids and other birds. Unfortunately, the recent addition of an entrance gate has placed this area off-limits to birders.

Farther north is the Puʻu Anahulu State Game Management Area, marked by an access road and hunter checking box on the east side of the highway at the 17.8-mile point. A State permit is required to enter this area on weekdays, while on weekends and holidays it is open to hunters and hikers. Birdlife is not particularly abundant here, but it is the best spot in this area to get away from the highway. Look for Saffron Finch, Yellow-fronted Canary, Northern Cardinal, Japanese White-eye, Warbling Silverbill, Nutmeg Mannikin, and Hawaiʻi ʻAmakihi.

As you drive along the highway between Kailua-Kona and Waimea, there are many other native and introduced birds to watch for. Native Pueo and endemic ʻIo sometimes soar overhead. At dusk you may see the introduced Barn Owl as well. Three introduced francolin species—Gray Francolin, Black Francolin, and Erckel's Francolin—may be found in the area. This is the only spot in the State that can boast four dove species. Spotted Dove, Zebra Dove, and Rock Dove are uncommon, and the Mourning Dove is established here in very small numbers. Nēnē and Chestnut-bellied Sandgrouse are sometimes seen flying upslope in the morning and back down in the evening. Wild Turkey, Common Peafowl, and Ring-necked Pheasant are other birds to watch for.

One other site along this route is worth mentioning even though it is not yet open to the public. The State recently established the Pu'u Wa'awa'a Wildlife Sanctuary on the slopes of Hualālai specifically to protect habitat for the endangered 'Alalā. Only a single crow has inhabited the area in recent years. The tiny remaining population of about twenty birds is farther south, on private ranch land above Kona. Successful captive breeding and rearing efforts are under way, and it is hoped that breeding 'Alalā will someday be reestablished on Hualālai.

Other Big Island Birding Spots

'Akaka Falls State Park

This lush park fits the popular image of Hawaiian rain forest—towering palms, bamboo and other exotic plants, a profusion of flowers, and waterfalls plunging hundreds of feet into deep pools. Nearly all the plants are introduced species, and so are the birds that flit among the treetops. While you stroll the loop trail watch for Northern Cardinal, Japanese White-eye, and House Finch. Don't be fooled by these finches. They are quite yellow, not the commoner red type usually seen. Around the parking area, look for mongooses begging from the picnickers.

To get to 'Akaka Falls, go north from Hilo about ten miles. Prominent signs will direct you inland on Route 220 four miles to the park at the end of the road. Bring your mosquito repellent if you plan to stand still for more than a few minutes. Also, do not leave valuables in your car, since break-ins have occurred at the parking area. Access is poor for the disabled birder. The trail is paved but there are hundreds of steps as the path descends into the gulch carved out by Kolekole Stream. Even the rest rooms do not have good access.

Laupāhoehoe County Beach Park

This area, along the Hāmākua Coast north of Hilo, is one of the few places on the Big Island where you can get a good vantage point to see marine birds. It is 16.9 miles north of Hilo, with a sign marking the turnoff. After leaving Highway 19 you descend 1.2 miles down a paved but narrow and winding cliff road to the peninsula where the park is located. This scenic area would be worth a visit even if there were no birds to be seen. (And sometimes there are no seabirds present, or they are flying so far out that identification is impossible.) On the drive down, watch for Nutmeg Mannikin, Zebra Dove, Spotted Dove, and Northern Cardinal. There are a few homes in the area, but most of the village of Laupāhoehoe was moved up the hill in 1946 after a devastating tsunami hit this peninsula. As you drive past the old homes that remain, watch for dogs sleeping in the road. This area doesn't get much traffic!

The county park is a fine one, with rest rooms at the north and south ends (the ones at the north end have better wheelchair access, but all have narrow stalls), pavilions, faucets, a grassy camping area, and almost no tourists ever.

It is often windy or rainy at this park, so it is fortunate that you can get good birding views from your car. Look for Black Noddies and Wedge-tailed Shearwaters out from the peninsula and White-tailed Tropicbirds soaring high above the shoreline to the north or south.

South Point

The southernmost point of the Big Island is also the southernmost point in the United States, appropriately named South Point (the Hawaiian name is Ka Lae). Looking south from the weathered cliffs, there is nothing between you and Antarctica except 7,500 miles of water and a sprinkling of exotic South Pacific islands. This windswept spot is better known for its archaeology than its birding. South Point may have been the landfall for the first Polynesian explorers perhaps fifteen hundred years ago, and there are many old *heiau* and stone walls in the area. You can reach South Point more easily than the ancient voyagers, by driving south on South Point Road for about eleven miles from its junction with Highway 11, six miles west of Na'alehu.

Marine birds to look for include Wedge-tailed Shearwaters and Black Noddies. White-tailed Tropicbirds and Great Frigatebirds may also be seen. With luck, spotting a Laysan Albatross is even a possibility. Along the shoreline east of the cliffs, and in grassy areas, watch for wintering Kōlea, Ruddy Turnstones, and Sanderlings. The Eurasian Skylark may be seen along grassy roadsides. Very lucky birders who visit South Point in the winter may be rewarded with the sighting of a Bristle-thighed Curlew. There are very few records of this bird on the main Hawaiian Islands, but it was common enough for the early Hawaiians to have a name for it: Kioea.

Kehena

There are two spots on the Big Island where Black Noddies can be seen roosting and nesting along sea cliffs. Hōlei Sea Arch in Hawai'i Volcanoes National Park, described earlier in the chapter, is the more popular because a lot of visitors are in the national park anyway. The other viewing area is Kehena, worth mentioning because it has better access for disabled birders—the birds can be seen from the paved parking area. The Kehena birds are fairly close to Hōlei Sea Arch, but you face a long drive to see them since Madame Pele elected to sever the highway with lava flows between the two spots. The Kehena viewing area is a paved highway turnout at the 19.1-mile point on Highway 137. To reach it, drive south from the town of Kea'au on Highway 130. At Pāhoa, take Highway 132 around the eastern tip of the island and down to Kehena.

Directly below the turnout there is a black sand beach frequented by nudists. You may have to suffer laughs from passing residents who don't realize you are watching the birds. Sometimes birders have to make these sacrifices. Note that vandalism of autos is quite common here.

Kailua-Kona

Birders in the Kailua-Kona area may see some of the following species around hotels, condominiums, gardens, and roadsides: Common Myna, Zebra Dove, Spotted Dove, House Sparrow, Java Sparrow, House Finch, Northern Cardinal, Yellow-billed Cardinal, Japanese White-eye, Nutmeg Mannikin, and Saffron Finch. Less common birds include Yellow-fronted Canary, Warbling Silverbill, and Lavender Waxbill. A little birdseed scattered on lanai or patio is often well rewarded.

South Kohala

Visitors staying at the swank resorts of South Kohala may see Common Myna, Zebra Dove, Spotted Dove, House Sparrow, House Finch, Northern Cardinal, Yellow-billed Cardinal, Japanese White-eye, Northern Mockingbird, and Nutmeg Mannikin. Less common birds include Saffron Finch and Yellow-fronted Canary. Around golf courses look for Kōlea and Gray Francolin. Golf course ponds sometimes attract migrant waterfowl, including Northern Pintail, Northern Shoveler, and other less common migrants.

Hilo

Common urban birds in Hilo include Zebra Dove, Spotted Dove, Common Myna, House Sparrow, House Finch, Northern Cardinal, Japanese White-eye, and Nutmeg Mannikin. These same species can be found south of Hilo in the Puna District, on the way to the noddy viewing site at Kehena.

7
Maui

In 1866, Mark Twain was sent to Hawai'i by the California newspaper that employed him. His assignment was to pen colorful travel letters for the entertainment of the newspaper's readers back in Sacramento. After a long break in his correspondence, he wrote, "I went to Maui to stay a week and remained five. I had a jolly time. I would not have fooled away any of it writing letters under any consideration whatever."

You may notice that this chapter on Maui is a bit shorter than most in the book. It isn't because I was following Twain's example. When it comes to birding, Maui has fewer good spots that are accessible to the public than some of the other islands. In keeping with Maui's reputation for superlatives, though, this handful of birding spots may reward you with endemics you can't see on other islands, as well as sightings of very uncommon migrant or visitor species.

Topographically, Maui is very nearly two islands. It was formed from two distinct volcanoes as they emerged from the Pacific, and the eroded remnants are connected today by a low, narrow isthmus. The steep mountains of West Maui reach their summit at 5,788-foot Pu'u Kukui. These mountains support remnant populations of several endemic bird species as well as rare bog ecosystems. The birds and bogs are nearly inaccessible in the steep terrain, and the private land is not open to the public. Fortunately, biologically significant areas of West Maui are being protected by The Nature Conservancy, working in cooperation with the landowner. East Maui is dominated by Haleakalā, the largest dormant volcano in the world. From its 10,023-foot summit at Pu'u 'Ula'ula you can scan the enormous multihued Haleakalā Crater, or look southeast over eighty miles to the even taller peaks of Mauna Kea and Mauna Loa on the Big Island.

Like the mountains to the south, Haleakalā provides refuge for endemic birds on its high, cool slopes. The rain forests of the windward face of the mountain have yielded some truly wondrous surprises. In 1973 the scientific community was astounded when two young researchers working in the forest discovered a bird entirely unknown to science, a little brown bird so rare, quiet, and unobtrusive that apparently it was also unknown to the early Hawaiians. This endangered bird, given the name Po'ouli, enjoys the protection of Haleakalā National Park land that is closed to the public. Its avian neighbors on this closed and protected mountain slope include four or five other endangered endemic forms as well. One of these may be the Bishop's 'Ō'ō. Believed extinct since the turn of the century, this long-tailed black honeyeater may have been spotted here in 1981. Although part of the mountain is closed, there are excellent birding areas open to the public in the national park, an adjacent Nature Conservancy preserve, and a state park.

The most popular lodging choices for Maui visitors line the dry,

Map 5
Maui Birding Sites

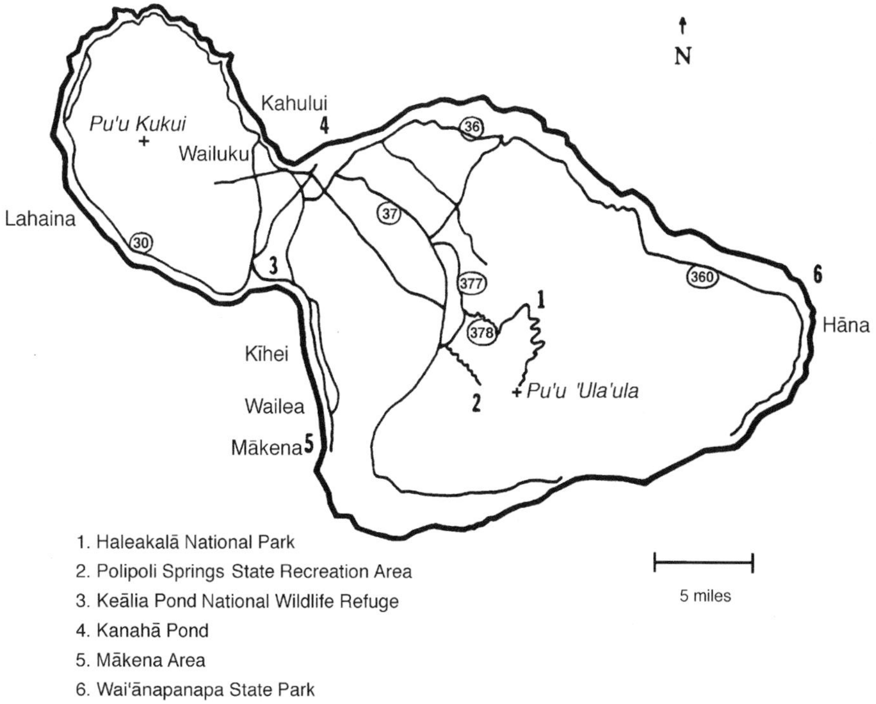
N
Kahului
4
Pu'u Kukui
Wailuku
Lahaina
30
36
37
3
377
360
6
1
378
Hāna
Kīhei
2
Pu'u 'Ula'ula
Wailea
Mākena 5
5 miles
1. Haleakalā National Park
2. Polipoli Springs State Recreation Area
3. Keālia Pond National Wildlife Refuge
4. Kanahā Pond
5. Mākena Area
6. Wai'ānapanapa State Park

sunny west coast. They include Lahaina and the resorts of Kā'anapali and Kapalua up on West Maui, and Kīhei and the Wailea area resorts to the south. All of these areas offer some viewing of introduced birds. All are a fair drive from the birding areas on Haleakalā, but there are few lodging choices that are any closer. Some visitors enjoy the lush green tropical setting at Hāna on the east side of the island. Although there is some notable birding in the area, Hāna cannot match the variety of birds found on other parts of the island.

For those who want to rough it, the State has one extremely rustic cabin for rent at Polipoli Springs State Recreation Area, in the midst of good birding and extensive hiking. During wet weather you need a 4WD vehicle to reach the cabin. Except for a wet-weather trip to Polipoli, there is no need for birders to have 4WD on Maui.

1. Haleakalā National Park

FEATURES:
- Many endemic birds
- Native plants
- Splendid views

FACILITIES:

Haleakalā National Park contains one of the most remarkable geologic features in the nation—a dormant volcano whose eroded crater is so vast and its rocks so colorful that visitors doubt their own perceptions. Birders may also doubt their perceptions when they see and hear the wondrous avian riches of this park. The steep drive up to Haleakalā almost always includes sightings of Pueo. 'Apapane and 'I'iwi are sometimes so plentiful at Hosmer Grove just inside the park entrance that you would say they are abundant. A couple of Hawai'i 'Amakihi at Hosmer are so adaptable that they have taken to mooching the tourists for food scraps. Nēnē mingle with visitors at park headquarters and crater overlooks. Extremely lucky birders on guided walks through the adjacent Waikamoi Preserve have spotted 'Ākohekohe and Maui Parrotbill. Visitors to the crater overlooks at night may hear the call of the Dark-rumped Petrel.

When you visit Haleakalā remember that you are at very high elevations. The trailhead at Hosmer Grove is the lowest on Haleakalā at 6,800 feet. Other birding spots and trails in the park are much higher. There is less oxygen than at sea level and you will feel the difference when hiking. Since there is also less atmosphere to screen the direct rays of the tropical sun, you will want sunscreen and a hat even on overcast days. Early in the morning or during cloudy and wet weather it can be very cold, and every few years the mountain is even dusted with snow. Most park restrooms have good disabled access.

To reach Haleakalā from Kahului, take Highway 37 to Highway 377 to Highway 378. This sounds confusing, but the way is well marked; Haleakalā is the most popular site on the island. As you drive up, up, up toward the national park, watch for Pueo soaring and hovering over the grasslands. This bird is a subspecies of the Short-eared Owl, which can be found on every continent except Australia. The Short-eared Owl is more visible than some other owls because it is active at dawn, dusk, and even at midday. The habit of hunting during daylight hours seems even more pronounced in the Pueo, making the bird a common sight along the road to Haleakalā at any time of day. Eurasian Skylarks and Ring-necked Pheasants are commonly seen along the roadsides. Occasionally Chukar can be seen as well, especially in rocky areas.

Don't get too carried away looking for birds as you drive, because there are a lot of hazards along this road: switchbacks that will make you feel like a Grand Prix driver, bicycle tours cruising down the hill, and cattle on the road. This is open range, pardner!

Hosmer Grove/Waikamoi Preserve

One of the better places in all Hawai'i to see endemic forest birds is at Hosmer Grove in Haleakalā National Park and the adjacent Waikamoi Preserve of The Nature Conservancy. Hosmer Grove is just inside the park entrance, and is well marked. It is open at all times, and camping is permitted at the tiny and informal trailhead campground. There is a one-mile loop trail that is always open, with an interpretive brochure available in a box at the trailhead. The trail is not wheelchair accessible.

Many of the trees in Hosmer Grove are introduced species, planted by Ralph Hosmer in 1910 to test various trees for commercial timber production in Hawai'i. Hosmer found that the trees grew faster than their root systems, and they were often blown down in high winds. There are also many endemic and native plant species in the area, marked with labels along the loop trail. Look for 'ōhi'a and māmane, two of the trees favored by endemic honeycreepers. These plants bloom nearly all year, peaking in summer.

In the grassy areas along the entrance road to Hosmer Grove, watch for Ring-necked Pheasant and Kōlea. Around the small parking lot, species include 'Apapane, Hawai'i 'Amakihi, Maui 'Alauahio or Maui Creeper, Japanese White-eye, House Sparrow, House Finch, Spotted Dove, and Northern Cardinal. In a remarkable adaptation to available food sources, a pair of 'Amakihi is sometimes seen around the picnic pavilion begging for food. The birding is even better along the loop trail, especially at the gulch overlook. I have never failed to see 'I'iwi and 'Apapane from this spot. If you are very fortunate, you may glimpse the endangered Maui Parrotbill. Lucky and observant visitors see this bird at the gulch on rare occasions. Other birds to watch for along the trail include Hwamei and Red-billed Leiothrix. Birders who have seen the endemic 'Elepaio on O'ahu, Kaua'i, or the Big Island might expect to see this bird at Hosmer Grove. Oddly, there is no evidence that the 'Elepaio ever established itself on Maui, Moloka'i, Lāna'i, or Ni'ihau.

The fences in this area were hastily erected to contain an outbreak of rabbits, an alien to Hawai'i whose establishment could be disastrous to native plants. Six rabbits were released illegally in the fall of 1989. Within a year, park rangers had trapped seventy-four of these fecund furballs and still hadn't eliminated them.

Adjacent to the Hosmer Grove area of the national park is Waikamoi Preserve, 5,230 acres of private land managed by The Nature Conservancy. The goal is to preserve and restore this area in order to protect the many species of endemic plants and animals that exist within the preserve boundaries. Fences are constructed to keep out goats and pigs, and volunteers remove invasive alien plants so that native vegetation can have a better chance to reestablish.

Participation in an organized tour of the preserve is the only way you can visit Waikamoi, and a visit is highly recommended. National Park rangers or volunteers lead birding walks along a two-mile trail in the preserve every Monday and Thursday. Hike days are subject to change, so be sure to check in advance. The Nature Conservancy also leads hikes regularly at Waikamoi. For information on preserve visits, contact Haleakalā National Park or The Nature Conservancy office in the nearby town of Makawao (see appendix).

The birding can be spectacular on these guided walks. You will very likely see all the common endemics mentioned above, and your guide will be able to help you sort out the little green birds you see: Hawai'i

'Amakihi, Maui 'Alauahio, and Japanese White-eye. On very rare occasions, 'Ākohekohe may be seen along the loop trail in the Waikamoi Preserve, although this endangered bird's usual habitat is farther east. The Maui Parrotbill has also been seen in this area a few times in recent years.

Many visitors to Maui drive to the top of the road on Haleakalā to watch the sunrise. Then, chilled to the bone, they drive back down the mountain and across the island to find a hot breakfast, as no food is available in the park. Here's a better idea. Take a thermos of hot coffee or tea and some trail breakfast with you. After sunrise at the summit, drive down to Hosmer Grove. A pavilion provides welcome shelter from wind and mist. Start a little fire in a barbecue pit (a stack of free firewood is usually provided but bring paper and matches to start a fire), warm up, have some breakfast, and then treat yourself to great birding. Wherever else you go, a visit to Hosmer Grove and the Waikamoi Preserve is a must for any birder visiting Maui!

Other Birding at Haleakalā

The endemic forest birds at Hosmer Grove and Waikamoi are dazzling, but they are not the only birding attractions at Haleakalā. The Nēnē is locally common and quite approachable in the park. This goose was driven to extinction on Maui but has been reintroduced on the mountain, and a population of about a hundred lives in the park. Nēnē can sometimes be seen around the park headquarters. If you want to see one that isn't standing on the front lawn of a National Park Service building, visit the Kalahaku Overlook or the Halemau'u trailhead to see Nēnē in a more natural setting. They are often seen in both places. Unfortunately, many of these geese have become beggars, taking handouts from visitors. This behavior makes it much harder for these birds to survive in the wild, and a diet of potato chips is not any healthier for birds than it is for people.

There is another bird at Haleakalā that you will not see but you may be lucky enough to hear. It is the Dark-rumped Petrel. The early Hawaiians knew this bird as the 'Ua'u, a name that echoes the sound of its call. The species breeds only in Hawai'i and the Galápagos Islands. At one time it was believed to be extinct in Hawai'i, but a small population was discovered nesting at elevations between 8,200 and 9,900 feet on the crater slopes of Haleakalā. There is evidence that the 'Ua'u once nested at much lower elevations, but it was apparently eaten by the pigs, dogs, and rats brought by the Polynesians, as well as the Polynesians themselves. Later introductions by Europeans of mongooses, cats, and disease-carrying mosquitoes nearly finished the birds off. Today this endangered species clings tenuously to existence in the most marginal of habitats, each pair trying to raise a single young in the cold and rarified atmosphere of the high mountain. In February and March these seabirds come to the crater to nest, in burrows that may be fifteen feet deep or more. Eggs are laid in April or May and hatch in July. The parents feed the young for nine weeks and then leave them. Within a few weeks, hunger drives the young birds to make the most dangerous flight of their lives, using untested wings to fly to the sea at night. The lucky ones are not confused by the lights of Maui, and make it.

If you visit Haleakalā from late April to August and come to Kalahaku Overlook about an hour after dusk, you might be lucky enough to hear these petrels calling. As you gaze across the vast moonlit crater and listen to the flutelike calls, perhaps you can imagine that you are in a Hawai'i never disturbed by the forces of change, the forces that have taken such a toll on the life of these Islands.

2. Polipoli Springs State Recreation Area

FEATURES:
• Jeepers! Abundant Creepers

FACILITIES:

Every Hawaiian island has, or had, a species of creeper. The creepers on O'ahu and the Big Island are endangered, the subspecies of creeper once found on Lāna'i is extinct, the Moloka'i Creeper is at or beyond the verge of extinction, and the Kaua'i Creeper is uncommon and restricted mostly to the Alaka'i Swamp. What's left? The Maui Creeper or Maui 'Alauahio— and it is abundant at Polipoli Springs State Recreation Area on the slopes of Haleakalā.

In 1993 the American Ornithologists' Union, considered the authority on naming of Hawaiian and North American birds, changed the common names of several Hawaiian creepers. At one time all the creepers were lumped into a single species, but in recent years separate species have been recognized on different islands. The Kaua'i and Big Island birds are even in a different genus from the others. It doesn't seem appropriate to call these widely varying birds creepers, so now most are known by their original Hawaiian names. The Hawai'i Creeper on the Big Island keeps its name, the Kaua'i Creeper is now the 'Akikiki, the Moloka'i Creeper is Kākāwahie, and the species on O'ahu is known as the O'ahu 'Alauahio. The Maui Creeper is now the Maui 'Alauahio.

Besides the satisfaction of large numbers of 'Alauahio, as well as other endemic, native, and introduced birds, your trip to Polipoli will be rewarded with breathtaking views (in clear weather), scenic trails through forests of introduced and some native trees, an opportunity to stay in a very rustic State-owned cabin, and a taste of the sweetest springwater anywhere. The only drawback of the area is that the unpaved road can be a bit of a challenge for passenger cars in good weather, and strictly 4WD when it's wet.

To reach Polipoli take Highway 37, the Haleakalā Highway, from Kahului. Just past the town of Pukalani, Highway 377 splits away from Highway 37 toward Haleakalā, runs parallel to 37 for a few miles, and joins 37 again. Stay on 37 to the second junction with Highway 377 and proceed left on 377 for 0.3 miles. At this point turn right up a steep, unmarked paved road. Go 0.9 miles to a gate and sign for Polipoli, check your odometer, and follow the road in. From this gate the road zigzags upward across open cattle range for 5.1 miles before the pavement ends. After reaching this point, you will want to drive cautiously and turn back or walk if the road is too wet. At 8.4 miles from the open range gate the road forks, with a sign pointing right toward Polipoli. From this point onward the road improves, having a gravel surface. Finally, at 9.1 miles, you will reach the parking area at the end of the road.

The elevation at Polipoli is about six thousand feet, so it can be quite cool. It is often clear in the morning or late afternoon, but clouds can blanket the area during the middle of the day. Birders who visit the area in the morning and leave at midday or early afternoon may find themselves driving down the mountain through open range in clouds that reduce visibility to forty feet. Remember that cows on the road may be reluctant to yield the right-of-way.

Although there are some groves of native trees at Polipoli, most of the forest was planted during the 1920s and 30s by the State and the Civilian Conservation Corps. Some of the trees they planted include redwood, eucalyptus, cypress, ash, cedar, and pine species.

There is a network of trails throughout the park. You can hike a loop through the area *if* you find all the trail junctions and take all the correct branches where the trails fork. Without a good trail map you may want to follow a single trail and return the way you came, or hike with a knowledgeable companion. Excellent descriptions of Polipoli trails are included in a book entitled *Maui Trails—Walks, Strolls, and Treks on the Valley Isle.* See the list of additional reading at the back of this book.

The Polipoli Trail is a good one for birders. It starts at the lower end of the parking area and immediately passes through a dense stand of introduced conifers where Maui 'Alauahio can be seen. In fact, at Polipoli this bird is limited almost completely to these introduced trees. Farther on, the trail passes grassy areas and native trees.

Another route for birders is the Redwood Trail, which leads past an abandoned cabin to an area where fuchsias grow wild and attract 'Apapane. The trailhead is located just before you reach the parking area. The trail passes the park's rental cabin and then descends rather steeply. At about the 0.8-mile point the Redwood Trail intersects the Tie Trail. You will want to keep to the right. At about 1.4 miles you will reach a dilapidated old ranger's cabin, nearly 900 feet in elevation below the trailhead. A few steps past the abandoned cabin, turn right onto the Boundary Trail and walk a short distance to the fuchsias. Be very careful to observe your surroundings so that if you take a wrong turn you can retrace your steps and get out.

During the drive up to Polipoli, watch for Eurasian Skylark and Ring-necked Pheasant in the grasslands. Pueo may be seen overhead. When the road reaches open woodlands, you may see Hawai'i 'Amakihi. As the vegetation gets more dense there will be more Japanese White-eye and Red-billed Leiothrix. At the State Recreation Area look for 'Apapane in the native forests and Maui 'Alauahio in the introduced trees, where it seems to have adapted very successfully to creeping along the rough bark of the introduced conifers, searching for its diet of insects. Other birds to watch for in the forest include Hwamei, Northern Cardinal, and House Finch. 'I'iwi may be spotted, but they are less numerous than farther to the northwest at Hosmer Grove.

Be sure to taste the water at the picnic pavilion near the parking area. It comes from Polipoli Spring and lives up to its reputation for being pure, refreshing, and delicious.

The State has a single housekeeping cabin for rent at Polipoli. It is extremely old and rustic, has no electricity, and is difficult to keep clean when rain turns the nearby trails muddy. To rent it contact the Division of State Parks in Wailuku or Honolulu well in advance (see appendix).

3. Keālia Pond National Wildlife Refuge

FEATURES:
• Good place to see uncommon visitor birds

FACILITIES:

Despite Hawai'i's isolation, nonresident birds wander to the Islands with surprising frequency. Among the most common of these stragglers are gulls. No gull species breed in Hawai'i or elsewhere on tropical islands in the Pacific, for these birds are inhabitants of continents and coastal shelf areas. Occasionally, however, individuals of several species turn up in Hawai'i. Usually they are immature birds that are lost or carried off course in winter storms. These birds often end up at coastal ponds, and one of the best places in Hawai'i to spot such stragglers is Keālia Pond on Maui. Among the gulls most frequently seen are Laughing Gull and Ring-billed Gull. Others include Bonaparte's Gull, Herring Gull, Glaucous-winged Gull, and Franklin's Gull. Keālia Pond is also an important nesting area for Black-necked Stilts and Hawaiian Coots. To protect habitat for these birds, a 700-acre area was set aside in 1992 as the Keālia Pond National Wildlife Sanctuary.

Keālia Pond is on North Kīhei Road, or Highway 31, just north of the town of Kīhei. The area includes seasonal ponds between the highway and the ocean, and a much larger permanent pond on the inland side of the road. There aren't any visitor facilities at the refuge, but there is good viewing from the wide shoulders of the road. Be careful when you stop at Keālia because there is a lot of traffic on this road and it moves fast. There are turnouts with viewing access to the *mauka* or inland pond area at mile points 1.7 and 2.7. In the winter, the seasonal pond area forms along the ocean side of the highway between mile markers 1 and 2. When there is water in this *makai* pond, it may provide the best viewing because the birds won't be too distant from you. Large groups of Black-necked Stilts sometimes congregate between the highway and the ocean. This may also be the best area to look for stragglers.

Year-round residents of Keālia Pond include Black-necked Stilt,

Hawaiian Coot, Black-crowned Night-Heron, and Cattle Egret. Common migrants include Northern Pintail and Northern Shoveler, Kōlea, Wandering Tattler, Ruddy Turnstone, and Sanderling. Declines in duck numbers in recent years have made it harder to see any individuals of less common species. If you are lucky, you may see Lesser Scaup, American Wigeon, or Green-winged Teal. Rare wading birds that might be seen in winter include Bristle-thighed Curlew and Long-billed Dowitcher. In addition to the gulls, a straggler that turns up here occasionally is the Caspian Tern.

Hawaiian field guides do not include detailed descriptions of all the stragglers that might turn up at Keālia. A North American field guide can be very helpful in identification of the ducks, gulls, and other uncommon birds that you might spot during a wintertime visit.

4. Kanahā Pond

FEATURES:
• Endangered wetland birds

FACILITIES:

Kanahā Pond cannot match the variety of birds found at Keālia Pond, but it is notable for another reason. Kanahā is about as centrally located as a birding spot can be. In fact, this important wetland is right in the middle of Kahului, almost next door to the airport. The 143-acre pond area was established as a State wildlife sanctuary in 1959 to protect nesting habitat for the Hawaiian Coot and endemic race of Black-necked Stilt. Today urban development surrounds the peaceful sanctuary, but it continues to serve as an important habitat for two of Hawai'i's endangered wetland birds.

Visitors have two ways to visit the sanctuary. First, there is a public observation shelter just off Highway 396. This shelter is always open, there is adequate parking, and it has good access for the disabled. Coots and stilts are often seen quite close to this shelter. Look for the entrance on the north side of Highway 396 just east of its junction with Highway 36.

Kanahā Pond

It is also possible to walk through the sanctuary on a network of dirt roads. A free permit must be obtained in advance from the Hawai'i Department of Land and Natural Resources, Division of Forestry and Wildlife, at 54 High Street in Wailuku. (Highway 30 from Lahaina and Kā'anapali becomes High Street as it enters Wailuku; the State Building is at the corner of High and Main). Office hours are Monday through Friday from 8:00 A.M. to 4:00 P.M. You will be given a map of the sanctuary when you pick up your permit. The refuge is closed during breeding and nesting season, from April 1 through August 30.

Despite the surrounding city, Kanahā Pond usually has few if any human visitors. The serenity is interrupted only by aircraft flying in or out of the nearby airport. The birdlife is abundant. During your visit you are likely to see as many as forty Black-necked Stilts. Hawaiian Coots and Black-crowned Night-Herons also are common. Less common but easy to spot is the introduced Cattle Egret. Between August and April, Kōlea are present in the sanctuary, and during the winter other migratory species may be spotted. Ducks most likely to be seen include Northern Pintail and Northern Shoveler. Uncommon visitors may include Lesser Scaup, American Wigeon, and Green-winged Teal. The Mallards here are apparently feral birds. Other winter visitors include Ruddy Turnstone, Sanderling, and Wandering Tattler. The dry kiawe thickets around the pond support Gray Francolin, Northern Cardinal, Japanese White-eye, Common Myna, Zebra Dove, Spotted Dove, and Nutmeg Mannikin. Also watch out for the huge African snails that frequent the area.

The best time to visit the dirt roads of the sanctuary is early morning, when the sun will be behind you most of the time. Whether you visit the observation shelter or hike through the sanctuary, be sure to close the gates behind you. This helps keep predators such as dogs and cats out of this rather urban sanctuary.

5. Mākena Area

FEATURES:
- Many introduced birds
- Scenic kiawe forest
- Excellent beach

FACILITIES:

Just south of the bustle of Kīhei and the manicured affluence of Wailea, an older, slower, and quieter Maui still exists. The pressures of development are bound to continue their encroachment, but for now birders can enjoy a drive down a country lane, visit dry yet lush kiawe forests, swim at spectacular beaches, and see a host of introduced birds.

Drive south of Kīhei and Wailea on Wailea Alanui, the broad, winding boulevard that passes through the Wailea development. Just one mile past the entrance to the Maui Prince Hotel you will come to the entrance to Mākena Beach State Park, a wonderful beach with parking and rest rooms. But wait! Don't head for the water. The good birding is along the wide dirt path through the kiawe trees between the parking lot and the beach, and south along the beach where it borders kiawe forest. The best time to visit is very early in the morning before the crowds of beach visitors arrive.

In this area you may see one of the flashiest introduced birds in Hawai'i, the Red-crested Cardinal from South America. This bird is common in resort areas on some other islands, but on Maui it is harder to find. Northern Cardinals and Japanese White-eyes are abundant at Mākena. Other bird species you are likely to see are Common Myna, Zebra Dove, Spotted Dove, House Finch, House Sparrow, and Northern Mockingbird. Also watch for Nutmeg Mannikins. Flocks of these birds can be seen feeding on grass seeds. The Warbling Silverbill is much less common, but may join Mannikin flocks. You will probably hear the piercing repeated "cheater, cheater, cheater" call of the Gray Francolin. Look for this very wary bird around the edges of golf courses that border the road through Wailea, or in the kiawe thickets at Mākena.

The beach itself is a fine one, and there is good snorkeling around Pu'u 'Ōla'i or Red Hill, the peninsula at the north end of the beach. There is some unofficial camping in the area, and a trail over Red Hill leads to an unofficial clothing-optional beach. I have never encountered any problems in this area, but there are many signs warning you not to leave valuables in your car.

Beyond this beach, paved Wailea Alanui narrows and gradually turns to gravel. Passenger cars can easily make it as far as the 'Āhihi-Kīna'u Natural Area Reserve, an underwater reserve with good snorkeling and diving when seas are calm. If you snorkel in this area, be careful to avoid the abundant sea urchins. The isolation makes unattended rental cars tempting targets near the end of the road. It is best to leave no valuables in your car.

There is one more bird to watch for at Wailea. Several attempts have been made over the last century to introduce the Helmeted Guineafowl to Hawai'i. This large, plump, chickenlike bird, dark gray with white speckles on its body and a horny plate on its head, has never thrived in the Islands even though it does well in captivity and has been introduced successfully elsewhere in the world. It does not currently appear on checklists of Hawaiian birds. However, small flocks of Guineafowl have been spotted along the edges of golf courses and roads in Wailea. Perhaps the well-irrigated resort area will provide the right conditions for this bird to establish itself.

6. Wai'ānapanapa State Park

FEATURES:
• Black Noddies, other seabirds

FACILITIES:

Just north of Hāna is Wai'ānapanapa State Park, a place that offers many enticements for visitors to the windward coast of Maui. Most people come to the park for the rental cabins, camping area, hiking trails, and the watery caves steeped in Hawaiian legend. Wai'ānapanapa is also a good place to view seabirds—particularly the Black Noddy.

Just behind the state park office is Pa'iloa Bay, and in the bay is a high rocky islet. This islet, and to a lesser extent the cliffs on the shore of the bay, are Black Noddy nesting areas. These dark terns, named for the peculiar nodding habits they exhibit during breeding, feed closer to shore than many other seabirds in Hawai'i, so they are often visible flying within the bay or roosting on the islet. Park just south of the park headquarters and walk along the top of the cliffs that edge Pa'iloa Bay to view the birds. If you visit early in the morning you can see them flying out of small caves and crevices on the islet. You may be reminded of the tiny car at the circus from which troops of clowns emerge. Dozens of noddies keep pouring out of the same few tiny holes in the rock.

Watch overhead for Great Frigatebirds, and scan the ocean for Brown Boobies. Both of these birds roost in the area, although they are not known to breed along this coast. Brown Noddies may also be seen. The two noddies are easiest to distinguish during breeding season, when the Black Noddy shows orange feet. Brown Noddy feet are always black. Other birds you will probably see are Common Myna, Spotted Dove, Zebra Dove, Japanese White-eye, Northern Cardinal, House Sparrow, House Finch, and Nutmeg Mannikin.

Wai'ānapanapa has good access and amenities for the disabled. The Black Noddies can be viewed from the paved parking area near the caves a short way north of park headquarters. The rest rooms at the caves also have good access for the disabled.

Pa'iloa Bay

To stay in one of the twelve cabins at Wai'ānapanapa, make your reservations well in advance with the Division of State Parks in Wailuku or Honolulu (see appendix).

Other Maui Birding Spots

Waihe'e Ridge Trail

Birders thirsting to see endemic birds in the West Maui mountains might be tempted to hike the Waihe'e Ridge Trail up to 2,563-foot Lanilili Peak, a trail elevation gain of fifteen hundred feet. Even at this elevation you will really be too low to reach areas where population densities of endemics are significant. You will have a very slim chance of seeing an 'Apapane, Hawai'i 'Amakihi, or 'I'iwi. In the grasslands at the beginning of the trail, Kōlea are abundant from August to April, and flocks of Nutmeg Mannikins are common. Also watch for Zebra Dove, Spotted Dove, and Common Myna.

To reach the area, take Highway 340 north out of Wailuku. At the 6.9-mile point is the entrance to Camp Maluhia, a Boy Scout camp. Follow this entrance road 0.8 miles to a sign that marks the trailhead. The trailhead can also be reached by driving around the north end of West Maui. Older maps show an unpaved section of road, but it is all paved now. Caution is still necessary, however: some of the paved sections of this little-used road hug steep hillsides and are not nearly wide enough for two cars to pass. This side of West Maui directly faces the moisture-laden trade winds, so clouds and rain are quite common.

ʻĪao Valley State Park

There are lots of reasons to come to this state park. Of course, there is the famous ʻĪao Needle, the hard basaltic core of a volcano that has otherwise been eroded away by the streams draining this part of the West Maui mountains. Then there's history. The Hawaiians regarded this as a very spiritual place: ʻĪao means "sacred light" in Hawaiian. A grisly episode in Hawaiian history was played out here as well, when Kamehameha I trapped and slaughtered the defending army of the Maui king in the valley.

While you're taking in the scenery and reflecting on the history, watch for birds. You won't see any rare native species, but it is a scenic location for spotting Northern Cardinal, Common Myna, Spotted Dove, House Sparrow, and Japanese White-eye. You can get away from the crowds by descending the paved paths to the stream, then following the dirt path upstream for a mile or so. The stream can rise swiftly when it is raining in the mountains above you. If you notice any rise in the water, get out fast.

The paved path extends several hundred yards before stairs block access to the disabled. Rest rooms are accessible.

Kīpahulu District

If you're looking for paradise, the hike to Makahiku and Waimoku Falls may be about as close as you'll ever get. These spectacular falls, two hundred and four hundred feet high respectively, are in the Kīpahulu District of Haleakalā National Park. Protection of rare endemic forest birds was the primary reason for adding the Kīpahulu District to the national park. However, these birds are found well up in the mountains, usually above four thousand feet, in native forest. On the hike up to the base of Waimoku Falls you reach an elevation of only about sixteen hundred feet, and the hike travels through forests of introduced species such as guava and bamboo rather than native trees that are more likely to harbor endemic birds. The pristine area above the falls is closed to the public for the protection of the birds that live there. Introduced birds you are likely to see along the trail include Common Myna, Spotted Dove, Northern Cardinal, and Japanese White-eye.

Here are a few tips for the trail. Rain can move in suddenly, so be prepared for it. At the very least, carry a plastic bag for your camera and binoculars. Even if it does not rain where you are, it may be raining above you in the forest. Keep an eye on the stream and if the water level starts to rise, get out fast to avoid being caught in a flash flood. Also, be sure to bring plenty to drink. There is no drinking water available at the trailhead, and you will lose a lot of liquids on the hike because the first part is hot and sunny, and the elevation gain to Waimoku Falls is over nine hundred feet. The National Park Service sign at the trailhead estimates a hiking time of one and a half hours for the four-mile round-trip. Most people will need more hiking time than this, and will want to linger quite a while at the two beautiful waterfalls on the trail.

Kapalua and Kāʻanapali

There are many introduced birds to be seen in these resort areas. Around the golf course at Kapalua look for Common Myna, Kōlea, Red-crested Cardinal, Spotted Dove, Nutmeg Mannikin, and Gray Francolin.

At Kāʻanapali, Black-crowned Night-Herons can be very approachable around golf course ponds. Other birds to watch for include Northern Cardinal, Red-crested Cardinal, House Sparrow, House Finch, Common Myna, Japanese White-eye, Nutmeg Mannikin, and Kōlea. The Hyatt Regency Maui has a collection of many exotic birds including penguins, flamingoes, swans, macaws, and cranes. You do not need to be a guest of the hotel to take a guided bird tour of the grounds.

8
Moloka'i

MOLOKA'I IS ONE of the least-visited islands in the state, either by tourists or residents of the other islands. It has only one hotel that could be called a destination resort, and only a handful of other lodging choices for visitors. This relative lack of tourism has enabled Moloka'i to maintain a slower pace and a rural atmosphere. If you come to this island, the relaxed attitude will probably be the reason for your visit, because Moloka'i offers no unique attractions in the way of birds. Every species you are apt to see on Moloka'i can be seen on other islands. There are some endemic forest birds left but their numbers are dwindling, and it is almost impossible for visitors traveling on their own to reach the rain forest where these birds survive.

However, if you seek to escape from the crowds and resorts on other islands and you want a taste of rural Hawai'i, a few days on Moloka'i might be just the ticket. But remember that this island has a slower lifestyle and fewer amenities than other islands where tourism really drives the economy. As a guest, you should adapt yourself to the local rhythm of life, and don't expect to be pampered to the extent you would on other islands.

There will be birds to see no matter where you go on the island. Drive almost any road and you are likely to see Gray Francolins and Black Francolins along the shoulder. It seems like most families on Moloka'i own a horse, and these horses are frequently tethered along the sides of the roads to graze. As often as not, each horse will be accompanied by a small contingent of Cattle Egrets waiting for insects to be rousted from the grass. Be sure to look up, for the Pueo is sometimes seen soaring overhead, especially on the western half of the island. At dusk and dawn, you may see a Barn Owl toward the eastern half of the island. The Great Frigatebird also soars over Moloka'i, especially over the wide beaches at the west end.

A Cattle Egret waits for dinner

Map 6
Moloka'i Birding Sites

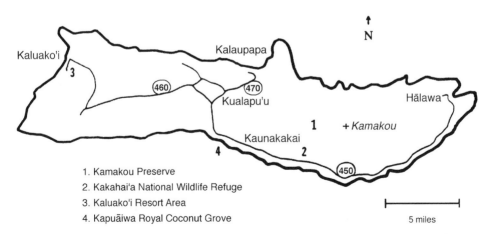

1. Kamakou Preserve
2. Kakahai'a National Wildlife Refuge
3. Kaluako'i Resort Area
4. Kapuāiwa Royal Coconut Grove

The island is thirty-eight miles long but only ten miles wide. This narrow strip of land has fewer mountains than the larger islands, and the highest peak, Kamakou, is only 4,970 feet high. The result is relatively little habitat suitable for endemic forest birds. Introduced animals and the activities of humans have taken their toll on the endemic flora and fauna of Moloka'i, too. Fortunately, some endemic species have found refuge in high and virtually inaccessible mountain areas.

Accommodations on Moloka'i include several choices at the west end, all within the Kaluako'i planned development, and several choices in or near the town of Kaunakakai. Either area has some fair birding, but Kaluako'i probably offers more species. The island is really small enough that you are never far from anything.

A budget alternative would be to camp in the bird-rich camping area at Pāpōhaku Beach Park near Kaluako'i. You will seldom have to share the place with more than one or two other groups, if any at all. A permit is required from the County Parks and Recreation Office (see appendix). Camping is also available at the cool uplands of Pālā'au State Park above the Kalaupapa Peninsula, north of Kaunakakai. A permit from the Division of State Parks is required. It can be issued by the State office on Maui (Moloka'i is part of Maui County) or the office in Honolulu (see appendix).

1. Kamakou Preserve

FEATURES:
• High elevation rain forest and bog
• Spectacular views

FACILITIES:

Although the mountains of Moloka'i are not as high or as extensive as those on some of the other islands, they are a stronghold for some of the world's rarest and most threatened plants, animals, and ecosystems. Much of the mountain area on the island is within the Moloka'i Forest Reserve, where the land is owned or controlled by the State of Hawai'i. Within the forest reserve, some areas receive even greater protection. The State has established two natural area reserves where unique and dwindling forest habitat is preserved and protected. Pu'u Ali'i and Oloku'i Natural Area Reserves protect plateaus between deep valleys on the north side of the island. These plateaus, especially Oloku'i, are nearly inaccessible to introduced animals that could damage the forest, and as a result they continue to harbor many rare plants.

Although the Natural Area Reserves are almost impossible to visit, there is some good access to the Moloka'i forest. In 1982, The Nature Conservancy acquired rights to protect and restore a 2,774-acre parcel in the heart of the State forest reserve. The Kamakou Preserve is home to over 250 plant species, more than 200 of them endemic to Hawai'i. The area is also home to several endemic birds. 'Apapane and Hawai'i 'Amakihi are fairly common, and a few 'I'iwi still survive in the forest. Oloma'o, a thrush closely related to the 'Ōma'o found on the Big Island, and Moloka'i Creeper are endemic to this island and are now critically endangered or extinct. The Oloma'o has not been seen since 1988, and no one has seen a Creeper on Moloka'i since 1963.

Management of Kamakou includes actions that reduce the impact of alien plants and animals. Fencing keeps destructive pigs at bay. Alien plants that encroach on the area are removed to the extent possible. The preserve receives a lot of rain, and the trails quickly turn into deep, muddy trenches. To protect the trails from erosion and to facilitate visits to the area, boardwalks have been constructed over some of the trails. These walkways can be extremely slippery when wet, but they are better than slogging through knee- or waist-deep mud.

The best way to visit the Moloka'i Forest Reserve area, especially on a first visit, is with a guided Nature Conservancy hike. These are held one Saturday each month and are scheduled a year in advance, so it is easy to plan a visit around one of these hikes. The convenient trips are set up so that you can fly in from another island for the day. The Nature Conservancy staff or volunteers will pick you up at the Moloka'i Airport in Conservancy 4WD vehicles and take you right to Kamakou Preserve. A modest donation is requested to help defray the cost of the trip. To reserve space, contact the preserve manager as far in advance as possible (see "Organized Hiking" in the appendix). These trips often fill up three or four months in advance.

The advantages of visiting Kamakou with Conservancy staff are that you will have someone knowledgeable to identify the bird calls and the amazing variety of endemic plants found in the area. Perhaps more important, you will enjoy the services of a skilled driver behind the wheel of the 4WD vehicle that takes you to the preserve. This is impor-

An organized hike at Kamakou

tant because there are few 4WD vehicles for rent on Moloka'i and the road to Kamakou is definitely not for the unskilled or faint of heart. The disadvantages of joining such a group are that you don't arrive at the preserve until after 10:00 in the morning, past the time of greatest bird activity, and you will be with a group of up to a dozen people who will be quite conspicuous to any nearby birds.

If you decide to visit the preserve on your own, be sure to contact the preserve manager in advance to check on the road conditions. Also, if you rent a 4WD vehicle make sure it really does have 4WD and that you are allowed to take it off the pavement. The front drive in some of the local rental vehicles has been permanently disengaged, and 4WD is absolutely essential to reach the preserve.

To reach the preserve, head west on Highway 460 out of Kaunakakai. Just before crossing the Manawainui Bridge, 3.7 miles from town, turn right (east) on the Forest Reserve Road. Soon the pavement ends, and the dust begins. A road branches to the right, leading to Red Hill. Keep to the left on the main road and watch for a sign reading "Molokai Forest Reserve." Nine miles from the highway you will pass Luanāmoku-'iliahi, the Sandalwood Measuring Pit. This boat-shaped hole in the ground was used to measure a shipload of sandalwood as it was harvested from

The view from Waikolu Lookout

the forests during the days when Hawaiian royalty sold this fragrant wood to traders for shipment to the Orient.

Another mile along the road brings you to the Waikolu Lookout, high above the lush Waikolu Valley. Some brave or foolish people come this far in passenger cars, but the quality of the road declines sharply past this point. From this overlook you may see 'Apapane or Hawai'i 'Amakihi in the 'ōhi'a trees below you. Also watch for White-tailed Tropicbirds soaring in the valley below. These birds have a three-foot wingspan but they look like white specks in the immensity of Waikolu Valley.

A few muddy, miserable miles later you reach a fork in the road. Keep to the left, and park at road's end just ahead. From this point you can follow a boardwalk over the mud through nearly untouched native forest. The Japanese Bush-Warbler is a very common resident. These birds are often heard but seldom seen. From September to December they halt their singing entirely, and you may not be aware of their presence. You are likely to see 'Apapane, and with luck you may see Hawai'i 'Amakihi.

Soon the boardwalk takes you across a bog where the native plants are stunted. At your feet is an 'ōhi'a forest just six inches tall. At the fork in the boardwalk, keep right and continue another half mile to reach the Pelekunu Valley Overlook. This valley is even larger than Waikolu. In 1988 The Nature Conservancy acquired control over nearly six thousand acres in the valley to protect aquatic organisms in Pelekunu Stream, one of the least-disturbed streams in the state. Look across this valley, if it is not obscured by clouds, to the Oloku'i Plateau on the other side. This area is protected on nearly all sides by steep valley walls and sheer ocean cliffs, and is so inaccessible that not even nimble goats and pigs have been able to reach the plateau in any great numbers. The result is a nearly pristine area where Hawaiian plants thrive, protected from the alien invaders brought by Polynesians and later by westerners.

2. Kakahai'a National Wildlife Refuge

FEATURES:

• Varied introduced birds

FACILITIES

Kakahai'a National Wildlife Refuge, located along the south shore of Moloka'i, supports a lot of wetland birds, and you can get very close to them. The only problem is that you can't see them. High bushes block all views from nearby Highway 450. But don't despair: there's a pleasant little beach park right across the highway, and plenty of introduced birds in the area. Nearby seasonal ponds may even harbor wetland birds in winter.

To reach the refuge, go east from Kaunakakai 5.4 miles. You will see Kakahai'a National Wildlife Refuge on the *mauka* side of the road and Kakahai'a Beach Park on the *makai* side. To see wetland birds, keep driving. Another wetland worth a look is just up the highway between the 9.0- and 9.3-mile points, on the *makai* side of the highway. This area has some open water during the winter and may attract migrants, and it is just visible from the road.

Along the way, watch for introduced birds. Many of the common species in the area prefer the drier hillsides above the road, but you may see some of them along the roadsides. Species to watch for include Kōlea, Gray Francolin, Black Francolin, California Quail, and flocks of Nutmeg Mannikins occasionally joined by Warbling Silverbills. Also present are Zebra Dove, Spotted Dove, Common Myna, House Sparrow, and House Finch.

3. Kaluako'i Resort Area

FEATURES:
• Many introduced and native birds

FACILITIES:

One of the best birding areas on Moloka'i is around the Kaluako'i Resort at the western tip of the island. Here you will find birds that frequent open arid land, birds that are attracted to developed areas and irrigated areas, and shorebirds along the best beaches on Moloka'i. The area is easy to find. Drive west on Highway 460 from Kaunakakai, past the airport. When you get to well-marked Kaluako'i Road, follow it down to the resort.

On the resort grounds look for House Sparrow, House Finch, Spotted Dove, Zebra Dove, Northern Cardinal, Red-crested Cardinal, Nutmeg Mannikin, Common Myna, and Northern Mockingbird. Kōlea are common from August to April. Along the edges of the golf course you may see any of these, plus Gray Francolin, Black Francolin, or even Wild Turkey. Occasionally Hawaiian Coots may be attracted to the golf course lakes. Along the beach in winter you may see Wandering Tattler, Ruddy Turnstone, and Sanderling.

A good place to look for all these birds except the Coot is Pāpōhaku Beach Park. This nice area is 1.1 miles past the Kaluako'i Resort entrance drive, or nearly six miles from the turnoff on Highway 460. Red-crested Cardinals are unusually abundant at Pāpōhaku. Flocks of Ruddy Turnstones can sometimes be seen feeding along the beach or even in the lawn area of the park during the winter season when they are found in Hawai'i. Both Gray Francolins and Black Francolins are quite common along the mowed roadsides. If you are lucky you may see Wild Turkeys in or around the park.

There is camping at the park (by County permit) in a level, shady kiawe grove, with rest rooms and drinking water. The rest rooms are accessible to the disabled, but the path to the building is a bit rough and has a two-inch step. Many birds can be seen from the parking lot.

The grasslands and woodlands of this area are good places to spot the introduced axis deer, native to India and Ceylon.

4. Kapuāiwa Royal Coconut Grove

FEATURES:

- Black-necked Stilts
- Historic coconut grove

FACILITIES:

A visit to the Kapuāiwa Coconut Grove and the adjacent Kioea County Beach Park will give you an opportunity to do some birding, appreciate Hawaiian history, and take in a magnificent sunset. To reach the area, drive west from Kaunakakai for 1.3 miles. The imposing coconut grove is obvious, but the park on the town side of the grove has no sign. Black-necked Stilts often congregate in the shallows just offshore when the tide is out, especially along the stretch of shoreline between the park and Kaunakakai Wharf. Low tides at sunrise or sunset seem to be the best time to see stilts. You can drive right to the shore in the county park. Look left for stilts, look right for the sunset.

The coconut grove was planted as a symbolic insurance policy for Kamehameha V, so that he would never worry about going hungry. Personally, I would starve long before I could ever get up one of those lofty trees to retrieve a fresh coconut. And speaking of food, after sunset you may see people wading in the water with lanterns suspended from floating inner tubes. They are catching crabs.

Other birds you are likely to see include the usual cast of urban characters, such as House Finch, House Sparrow, Zebra Dove, Spotted Dove, Common Myna, Red-crested Cardinal, and perhaps Kōlea on the lawn.

The beaches all along the south shore of Moloka'i, especially below the mountains on the eastern half of the island, are muddy or silty rather than sandy. This is a legacy of the imprudent woodcutting and grazing of the island's past. The silt has eroded from the mountains and buried most of the beaches and corals that occurred naturally.

Other Moloka'i Birding Spots

Pālā'au State Park

The north coast of Moloka'i has the tallest sea cliffs in the world. For the most part, these cliffs are toward the inaccessible eastern side of the island, but you can get an idea of their grandeur at Pālā'au State Park. The clifftop lookout also provides an excellent bird's-eye view of the Kalau-papa Peninsula far below. To get to the park take Highway 460 east, then north out of Kaunakakai and keep to the right on Highway 470 where the road splits. The park is at the end of the road, past the stables that offer mule excursions to Kalaupapa National Historical Park.

On the way to Pālā'au watch for Gray Francolins and Black Francolins along the roads. In groves of ironwood trees at the park look for Japanese White-eyes. At the lookout you may see White-tailed Tropic-birds sailing far below you. Although you are almost a third of a mile above sea level, with luck you may spot sea turtles or dolphins in the water below. It is usually cool at Pālā'au, even when it's hot in Kaunaka-kai, because of the 1,600-foot elevation and the shade of introduced iron-wood trees that abound in the park. There are picnic tables and a woodsy but seldom-used campground. The water here is not potable, and access for the disabled is poor.

The view from Pālā'au State Park

Moa'ula Falls

Near the eastern tip of Moloka'i is Hālawa Valley, and at its head, Moa'ula Falls offer a popular hiking destination for visitors. There are relatively few birds in the forest of mostly introduced trees along the trail, but you will probably see Japanese White-eye, Spotted Dove, and Common Myna.

The 2.5-mile trail climbs only about 250 feet, making it a fairly easy walk. The route passes the ruins of many old home sites and taro patches now overgrown with huge trees, giving the area the feel of some ancient enchanted forest. Since the trail crosses the stream, you will want to have nonslip, waterproof footwear such as surfers' aquatic socks. If you try to hop across the boulders or wade through barefoot, you could easily slip.

'Ō'ō'ia Pond and Kaluaapuhi Pond

The south coast of Moloka'i is riddled with dozens of ancient fishponds used by the early Hawaiians for aquaculture. Most of them jut into the ocean and are unprotected, but these two ponds just west of Kaunakakai are surrounded by introduced mangroves and provide some habitat for wetland birds. They are hard to find and offer little viewing because they are on private land.

To reach the area, the adventurous birder can head west from Kaunakakai on Highway 460. At the 1.9-mile point is Kahanu Avenue. Follow it south toward the ocean and the town dump. When you reach an intersection and the dump entrance, turn and head west on this dump road. Alternatively, go south from the highway at the 2.7-mile point, at the Ocean Research Center. You merge with the dump road. Continue southwest one mile to a point where you can see the open water of Kaluaapuhi Pond. Disregard the abandoned collapsing buildings, and look for Hawaiian Coots, ducks, or other wetland birds that may be visible from the road.

9
Lāna'i

The history of Hawai'i is a story of radical changes for the people and the fragile environment. Today, the forces of change are hard at work on Lāna'i. Until just a few years ago this island was a sleepy time capsule, a 14,000-acre pineapple farm surrounded by mountains and grasslands on an island where 96 percent of the land is owned by a single corporation. Now that corporation has decided that pineapples can be grown more cheaply abroad, and pineapple culture is being replaced by lavish destination resorts. The transformation, however, is far from complete, and Lāna'i still retains much of the charm of an earlier time.

Lāna'i doesn't have the tall mountains of its loftier island neighbors, with the summit of Lāna'ihale reaching only to 3,370 feet. The native forests faced the same overgrazing as on the other islands, and they have been greatly diminished. Plantings of introduced trees include many Norfolk Island pine, giving the cloudy upper slopes of Lāna'ihale a woodsy—but not particulary tropical—feel.

Sadly, the endemic birds of Lāna'i are almost totally gone. There are perhaps a few hundred 'Apapane left, and these may be surviving only through periodic natural infusions of birds from Moloka'i or Maui. The Hawai'i 'Amakihi may still survive in low numbers, but has not been seen since 1976. Still, if you come to Lāna'i to catch a glimpse of the slow rural life or to be pampered at one of the luxury resorts now operating, there are birds for you to see. Introduced game birds abound. Other introduced birds can be seen around the island's many historical sites, in the fields, and at the beach.

There are only three commercial lodging choices on the island. The old Lāna'i Hotel in Lāna'i City was built in 1923 as a guest house for the pineapple plantation. Its eleven rooms are adequate and less expensive than other lodging on the island. The Lodge at Kō'ele and the Mānele Bay Hotel are the new accommodations located on the slopes of Lāna'ihale and at Hulopoʻe Bay, respectively. These resorts offer interpretive programs and island tours for their guests but no programs that emphasize the island birdlife.

There are only twenty-two miles of paved road on the island. Your exploring will be limited without a 4WD vehicle. They are for rent but it is a good idea to reserve in advance. Also helpful for exploring is the "Road and Site Map" of the island, available free from the car rental agency or the hotels.

Map 7
Lānaʻi Birding Sites

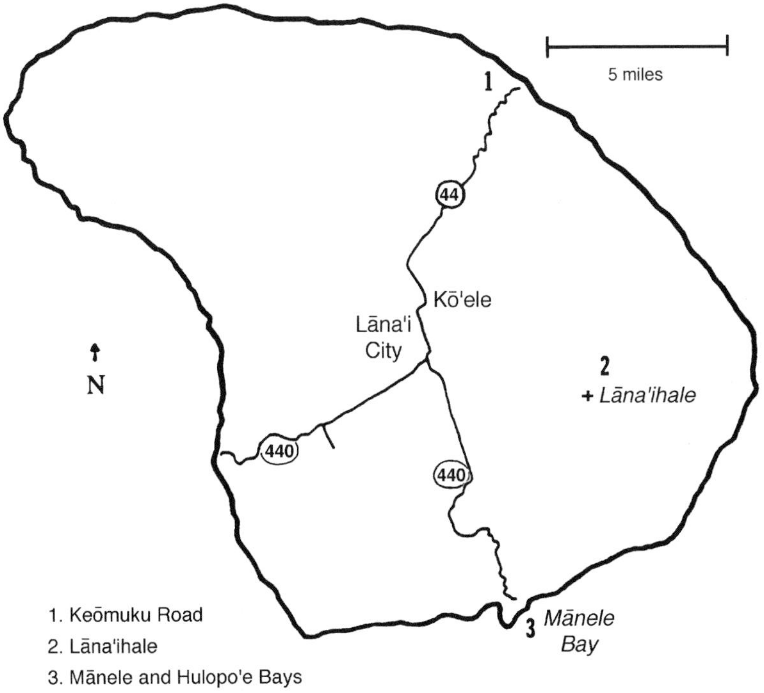
5 miles
1
44
Kō'ele
Lāna'i City
N
2
+ Lāna'ihale
440
440
3
Mānele Bay
1. Keōmuku Road
2. Lāna'ihale
3. Mānele and Hulopo'e Bays

1. Keōmuku Road

FEATURES:

• Petroglyphs
• A shipwreck
• Wild Turkeys

FACILITIES:

Some good dry-country birding, great views, ancient petroglyphs, and a shipwreck combine to make a trip down Keōmuku Road an exciting morning's outing. To reach the area, take Lāna'i Avenue north out of Lāna'i City. At the edge of town, where the road passes the golf course and the Kō'ele Lodge, it becomes Keōmuku Road. Follow it north for about eight miles, over the ridge and down toward the sea. The road is paved all the way down to the flat, kiawe-covered coastal plain.

At the end of the pavement, the road forks. One fork continues straight, then veers right (southeast) as it approaches the coastline. This road leads to the old townsite of Keōmuku, which is six miles down the coast. There is some worthwhile birding near the beginning of this level road as it passes through a shady kiawe grove, and the walking is easy. You will see many bird tracks in the fine sand. The huge Wild Turkey tracks are easy to spot. With any luck, you will see a few of these birds making tracks—they are fairly common in this area. Other birds to watch for include Erckel's Francolin, Gray Francolin, Zebra Dove, Common Myna, European Skylark, Northern Mockingbird, Nutmeg Mannikin, and Warbling Silverbill. Northern Cardinals are abundant, and there are a few Red-crested Cardinals. There are a few Japanese White-eyes in the area, but they seem far less common on Lāna'i than on other islands. Along this coastline are several old shacks used by local people, mainly on weekends. Chances are you will have the area all to yourself on a weekday.

Back where the pavement ends, take the left fork to see the shipwreck and petroglyphs. This road extends for about one and a half miles, most of it passable in a passenger car. It can get hot along the road because it is not well shaded by the kiawe, so come early and bring plenty of water.

At the end of the road is a turnaround, a primitive picnic area, and a concrete foundation that once supported a beacon. From this spot you can look up the coast to the rusted wreck of a large ship that has been stranded on the reef for about fifty years.

From the beacon foundation follow the path marked by stone cairns and old daubs of white paint inland for about five minutes. There, in the gully of a dry streambed shaded by kiawe, are some very nice petroglyphs. This is a good place to have a cool drink, listen to the doves and the wind in the kiawe branches, and wonder at the artists who left their marks here so long ago.

2. Lāna'ihale

FEATURES:
- Chance to see 'Apapane
- Good views in clear weather

FACILITIES:

If you want to see an endemic forest bird on Lāna'i, then the remnant native forest on the slopes of Lāna'ihale is the place to go. Even up on the mountain, the only endemic known to survive is the 'Apapane, and it is uncommon. To reach the area, drive north in your 4WD vehicle on Keōmuku Road from Lāna'i City about a mile toward the trailhead, a paved road heading right (southeast) from the highway. It is not marked, but it is the only paved road branching from the highway in this area. It is shown clearly on the free "Road and Site Map" as the Munro Trail.

The pavement ends a short distance ahead, near an old cemetery. From here the dirt road is fairly good as 4WD roads go. However, it should only be attempted in dry weather, which can be rare on this mountain. If it has rained recently, or if the Norfolk Island pines have been collecting fog and dripping it down on the road, it will be

Lāna'ihale

extremely slippery. Your tire treads will fill up with mud and your vehicle will dance the hula on the steep slope. It is a long way to the bottom if you slide over the edge.

Two introduced birds you may see are Japanese White-eye and Northern Cardinal. To the hopeful birder anxious to see endemic species, these could look like Hawai'i 'Amakihi and 'Apapane. Be sure to take a close look. The most prominent evidence of birdlife will probably be the shy Japanese Bush-Warbler, which you will hear but not see.

If you make it to the top, you can return the way you came or continue south, into the abandoned pineapple fields, and make your way back to paved Mānele Road. In the fields watch for Ring-necked Pheasant, Gray Francolin, and Erckel's Francolin. Overhead you may see Pueo. Where there is tall grass with seed heads, watch for Nutmeg Mannikin and Warbling Silverbill.

You can also walk a portion of the Munro Trail from the Lodge at Kō'ele. An excellent brochure describing the Koloiki Ridge Hike is available from the hotel concierge. As you start your walk from the Lodge, watch for Wild Turkeys. I am told that a flock has been frequenting the grounds.

3. Mānele and Hulopo'e Bays

FEATURES:
• Nice beach park
• Introduced birds

FACILITIES:

There are only three paved roads leading out of Lāna'i City. One of them goes to the twin bays of Mānele and Hulopo'e on Lāna'i's south coast. At Mānele Bay is the island's public boat harbor. Nearby at Hulopo'e Bay is a beach park and a State marine life conservation area. Hulopo'e is popular with local picnickers, swimmers, snorkelers, and surfers. Hulopo'e Bay is also the site of the Mānele Bay Hotel, largest on the island.

Along the road to the bays, watch for Pueo over the fields. Also watch for Ring-necked Pheasants, Gray Francolins, and Erckel's Francolins along the road shoulders.

At the bays the birds are nearly all introduced species, and most take advantage of handouts from park users. Look for abundant Northern Cardinals and a few Red-crested Cardinals. Other species include Common Myna, House Sparrow, Northern Mockingbird, Spotted Dove, Zebra Dove, and Nutmeg Mannikin. This area is home to a flock of about twenty Wild Turkeys. These turkeys are less wary than most of their species and supplement their diet with handouts. Some of these birds are very large. The one native bird that might be seen in this area is the Brown Booby. Sometimes this bird can be seen roosting on buoys at the boat harbor.

Other Lāna'i Birding Spots

Most of the birds mentioned in the site descriptions above can be seen at many places around the island. Ring-necked Pheasants, Gray Francolins, Erckel's Francolins, and Wild Turkeys are especially widespread and common.

Kaunolū

California Quail and the very similar Gambel's Quail are both found on Lāna'i, but they can be wary and hard to spot. Check the edges of fields and the rugged grasslands near the old Hawaiian village of Kaunolū at the southern tip of the island. The countryside around Kaunolū is also a good place to look for axis deer.

APPENDICES

Occurrence Tables

THE FOLLOWING TABLES will show you at a glance which birds you have a chance of seeing at each site on each island. In Hawai'i most birds are nonmigratory, but distribution and abundance can vary according to food supply. For instance, honeycreepers will follow the 'ōhi'a bloom and may be abundant or absent at a site depending on the condition of the trees. For the few migratory species, the tables show abundance in season. The symbols used to denote a bird's occurrence at a site are:

C: *Abundant to common.* A birder will have an excellent to fair chance of seeing this bird during most visits to a site.

U: *Uncommon to rare.* Present in much lower numbers compared to common species, or somewhat difficult to spot. May not be detected on every visit. Finding the rarest of birds in this category may require days of searching or uncommonly good luck.

Very rare birds that might be seen at a site only a few times per year or less are not included, nor are birds that occur on an island but only at inaccessible areas not described in this book. Some species that occur at the sites in this book are not usually included because they are very difficult to observe. Two examples are the Barn Owl, most likely to be out at dusk or later, and the Japanese Quail, which is tiny and very hard to flush out.

The field guide most likely to be used in conjunction with this book is *Hawaii's Birds* by the Hawai'i Audubon Society. For convenience, birds listed in the following tables appear in the same order as in the Audubon guide. Note that some birds described in *Hawaii's Birds* are hard to find and are not listed here.

The distribution and abundance of Hawaiian bird species are constantly changing. Over time, some birds will die out at some sites while other species will establish themselves. If you spot a new bird at a site described in this book, or if you visit a site regularly and find that a bird listed in these occurrence tables no longer occurs there, please let me know. Write to:

Rick Soehren
c/o University of Hawai'i Press
2840 Kolowalu Street
Honolulu, Hawai'i 96822

Oʻahu Birds

	Kapiʻolani Park	Diamond Head	Makiki Valley	Lyon Arboretum	Waʻahila Ridge	ʻAiea Loop Trail	Makapuʻu Point	Kawainui Marsh	Hoʻomaluhia Park	Kaneohe MCBH	Kualoa Park	Kahuku Area	Kaʻena Point	Sand Island	Hanauma Bay	Sacred Falls
Laysan Albatross										U			C			
Wedge-tailed Shearwater							C			C	C		U			
Red-tailed Tropicbird							U			U						
White-tailed Tropicbird							C			C	C		U			U
Red-footed Booby							C			C			C	U		
Brown Booby							U			C			U			
Great Frigatebird	U	U					C	U		C	C	C	C	U	C	
Sooty Tern							C			C						
Gray-backed Tern										U						
Black Noddy							C			U	C					
Brown Noddy							C			C			U			
White Tern	C	C					U									
Pomarine Jaeger													U			
Black-crowned Night-Heron								C	C	C	C	C				U
Koloa								U	U	U		C				
Fulvous Whistling-Duck												C				
Mallard								U	C	C		C				
Northern Pintail								U				C				
Northern Shoveler								U				C				
Hawaiian Coot								C	C	U	U	C				
Green-winged Teal												U				
Common Moorhen								U	C			C				
Black-necked Stilt								C	C	C	C	C				
Long-billed Dowitcher												U				
Kōlea	C	C	C	C			C	C	C	C	C	C		C	C	C

(continued)

O'ahu Birds

	Kapi'olani Park	Diamond Head	Makiki Valley	Lyon Arboretum	Wa'ahila Ridge	'Aiea Loop Trail	Makapu'u Point	Kawainui Marsh	Ho'omaluhia Park	Kaneohe MCBH	Kualoa Park	Kahuku Area	Ka'ena Point	Sand Island	Hanauma Bay	Sacred Falls
Ruddy Turnstone							U			C	C	C	U	C		
Sanderling							U			U	U	C	U	U		
Wandering Tattler							C			U	U	C	U	U		
Sharp-tailed Sandpiper												U				
Pectoral Sandpiper												U				
Pueo												U				
Barn Owl	U															
Gray Francolin		U														
Ring-necked Pheasant									U		U					U
'Elepaio					U	U										
O'ahu 'Amakihi			U	U	C	C										
'Apapane					U	U	U									
Hwamei									U							
White-rumped Shama	U	U	C	C	C	C			C							
Japanese Bush-Warbler			U	U	U	C			U			C				C
Red-billed Leiothrix			U		U	U										
Cattle Egret								C		C	C	C			U	U
Rose-ringed Parakeet	U		U	U				U								
Red-crowned Amazon	U															
Japanese White-eye	C	C	C	C	C	C	U	C	C	C	U	U	U	C	C	C
Spotted Dove	C	C	C	C	C	C	C	C	C	C	C	C	C	C	C	C
Zebra Dove	C	C	C	C	C	C	C	C	C	C	C	U	U	C	C	C
Rock Dove	C	U	U				U					U			U	C
Red-vented Bulbul	C	U	C	C	C	C	C	U	C	C	C	U	C	C	C	C
Red-whiskered Bulbul	U		C	C	C				C							

(continued)

O'ahu Birds

	Kapi'olani Park	Diamond Head	Makiki Valley	Lyon Arboretum	Wa'ahila Ridge	'Aiea Loop Trail	Makapu'u Point	Kawainui Marsh	Ho'omaluhia Park	Kaneohe MCBH	Kualoa Park	Kahuku Area	Ka'ena Point	Sand Island	Hanauma Bay	Sacred Falls
Common Myna	C	C	C	C	U	U	C	C	C	C	C	C	C	C	C	C
Northern Mockingbird	U	C	U				C		U		U				U	
Northern Cardinal	U	U	C	C	C	C	U	U	C	C	C			C	C	C
Red-crested Cardinal	C	C	C	C		C	C	U	C	C	C			C	C	C
House Sparrow	C	C	C	C	C	U	C	C	C	C	C			C	C	C
House Finch	C	C	C	C	C	U	C	C	C	C	C	C		C	C	C
Yellow-fronted Canary	C	U							U							
Red Avadavat			U									C	C			
Common Waxbill	U	U	U		C	C			U		C	C				U
Orange-cheeked Waxbill		U							U							U
Java Sparrow	C	C	C	C					C	C			U	C	C	U
Nutmeg Mannikin	C	C	C	C	C	U	U	C	C	C	C	C		C	C	U
Chestnut Mannikin	U	U							U				U			

Kaua'i Birds

	Kōke'e/Alaka'i Area	Kīlauea Point	Hanalei NWR	Wailua River	Mānā Ponds	Kōloa Area	Kē'ē Beach	Kaua'i Marriott	Kukui o Lono Park	Salt Pond Beach
Laysan Albatross		C					U			
Wedge-tailed Shearwater		C					U			
Red-tailed Tropicbird		C								
White-tailed Tropicbird	C	C	U	C			C			
Red-footed Booby		C					C			
Brown Booby		U					C			
Great Frigatebird		C				U	C			
Brown Noddy							C			
Black Noddy							U			
Black-crowned Night-Heron			C	U	C	C				
Koloa			C	U	U	U			U	
Fulvous Whistling-Duck			U	U						
Northern Pintail			U		U	U				
Northern Shoveler						U				
Hawaiian Coot			C	C	C	C		C		
Common Moorhen			C		U			U		
Black-necked Stilt			C		C					U
Kōlea	C	C	C	C	C	C		C	C	C
Wandering Tattler		U	C				U			
Nēnē		U						C		
Pueo	U	U	U		U				U	U
Erckel's Francolin	C				U					
Ring-necked Pheasant	C	U	U	U	U	U				
Western Meadowlark			U	U			U			
'Elepaio	C									

(continued)

Kaua'i Birds

	Kōke'e/Alaka'i Area	Kīlauea Point	Hanalei NWR	Wailua River	Mānā Ponds	Kōloa Area	Kē'ē Beach	Kaua'i Marriott	Kukui o Lono Park	Salt Pond Beach
'Akeke'e	U									
'Anianiau	U									
Kaua'i 'Amakihi	C									
'Akikiki	U									
'Apapane	C									
'I'iwi	U									
Hwamei	C	U	C	C		C				
Greater N. Laughing-thrush					U	U				
White-rumped Shama	C	U	C	C	C	C	U	C	C	
Japanese Bush-Warbler	U		U	U			U			
Red Junglefowl	C			C		C			C	
Cattle Egret			C	C	C	C	U	C	U	U
Rose-ringed Parakeet									U	
Japanese White-eye	C	C	C	C	C	C	C	C	C	C
Spotted Dove	C	C	C	C	C	C	U	C	C	U
Zebra Dove	C	C	C	C	C	C	U	C	C	C
Common Myna	C	C	C	C	C	C	C	C	C	
Northern Mockingbird	U	U			U	U				C
Northern Cardinal	C	C	C	C	C	C	C	C	C	C
Red-crested Cardinal			C			C			C	U
House Sparrow		U		U		C		C	C	C
House Finch	C	C	C	U	U	C	C	C	U	U
Warbling Silverbill						U				
Java Sparrow							C			
Nutmeg Mannikin	C	C	C	C	C	C		C	C	C
Chestnut Mannikin	U			U		C		C	C	U

Hawai'i Birds (1)

	Kīpuka Puaulu	Mauna Loa Road	Thurston L. Tube	Kīpuka Nēnē	Hōlei Sea Arch	'Ōla'a Tract	Tree Planting Rd.	Kaumana Trail	Kīpuka 21	Pu'u 'Ō'ō/P.L. Rd.	Hakalau Area	Pu'u Huluhulu	Pu'u Lā'au
Laysan Albatross					U								
Wedge-tailed Shearwater					U								
White-tailed Tropicbird				C									
Great Frigatebird													
Black Noddy					C								
Black-crowned Night-Heron													
Koloa											U		
Pied-billed Grebe													
Northern Pintail											U		
Northern Shoveler											U		
Hawaiian Coot													
Black-necked Stilt													
Bristle-thighed Curlew					U								
Kōlea	C			C	C						U	U	C
Ruddy Turnstone	C										C		
Sanderling													
Wandering Tattler													
Nēnē				U	C				U	U			
'Io				U	U		U	U			U		
Pueo				U	U						U	U	U
Barn Owl													
Black Francolin													
Gray Francolin													U
Erckel's Francolin					C								U
Red-billed Francolin					U								

(continued)

	Kīpuka Puaulu	Mauna Loa Road	Thurston L. Tube	Kīpuka Nēnē	Hōlei Sea Arch	'Ōla'a Tract	Tree Planting Rd.	Kaumana Trail	Kīpuka 21	Pu'u 'Ō'ō/P.l. Rd.	Hakalau Area	Pu'u Huluhulu	Pu'u Lā'au
Chukar	U												U
California Quail	U		U							C			U
Gambel's Quail											U		
Japanese Quail											U		
Wild Turkey										U	U		U
Common Peafowl													
Ring-necked Pheasant	U	U		U						U	U		U
Chestnut-bellied Sandgrouse													
Eurasian Skylark		C									C		C
'Elepaio	C	U	C			C	U	C	C	C	C		U
'Ōma'o	C	U	C			C	C	C	C	C	C	U	
Palila													U
'Ākepa										U	U		
Hawai'i 'Amakihi	C	C	U	C		U	C	C	C	C	C	C	C
'Akiapōlā'au										U	U		U
Hawai'i Creeper										U	U		
'Apapane	C	C	C	U		C	C	C	C	C	C	C	U
'I'iwi		C				C	C	C	C	C	C		U
Hwamei	C	U	C			U	U				U		U
Red-billed Leiothrix	C	C	U					U	U	C	C		U
Kalij Pheasant	C	C	U			U		C	C	U	U		U
Cattle Egret													
Japanese White-eye	C	C	C	C		C	C	C	C	C	C	C	C
Spotted Dove	U	U	U	U							U		U
Zebra Dove	U	U	U	U							U		U

(continued)

Hawai'i Birds (1)

	Kīpuka Puaulu	Mauna Loa Road	Thurston L. Tube	Kīpuka Nēnē	Hōlei Sea Arch	ʻOlaʻa Tract	Tree Planting Rd.	Kaumana Trail	Kīpuka 21	Puʻu ʻŌʻō/P.L. Rd.	Hakalau Area	Puʻu Huluhulu	Puʻu Lāʻau
Mourning Dove													
Rock Dove													U
Common Myna	U	U	U	U					U		U	U	U
Northern Mockingbird										U		U	U
Northern Cardinal	C	C	C	U		U	C				U		U
Yellow-billed Cardinal													
House Sparrow			U								U		U
House Finch	C	C	U	U		U		C	C	C	C	C	C
Saffron Finch													
Yellow-fronted Canary													U
Red Avadavat													
Warbling Silverbill												U	U
Java Sparrow													
Nutmeg Mannikin	U	C											
Lavender Waxbill													

Hawaiʻi Birds (2)

	Manukā	Whittington	Loko Waka Pond	ʻAimakapā Pond	Kaloko Drive	Puʻu Anahulu	ʻAkaka Falls	Laupāhoehoe	South Point	Kehena	Kailua-Kona	Kohala Resorts	Hilo Town
Laysan Albatross								U	U				
Wedge-tailed Shearwater			U					U	U				
White-tailed Tropicbird			C					C	U				
Great Frigatebird									U				
Black Noddy								U	U	C			
Black-crowned Night-Heron		C	C	C								U	
Koloa												U	
Pied-billed Grebe				C									
Northern Pintail			U	U								U	
Northern Shoveler			U	U								U	
Hawaiian Coot			C	C									
Black-necked Stilt				C									
Bristle-thighed Curlew									U				
Kōlea	C	C	C		C				C		C	C	C
Ruddy Turnstone		C		C					C	C		U	
Sanderling				U					U			U	
Wandering Tattler		C		C				C	U			U	
Nēnē						U							
ʻIo					U		U	U					
Pueo						U		U					
Barn Owl						U						U	
Black Francolin			U	U									
Gray Francolin						U						C	
Erckel's Francolin						C							
Red-billed Francolin													

(continued)

	Manukā	Whittington	Loko Waka Pond	'Aimakapā Pond	Kaloko Drive	Pu'u Anahulu	'Akaka Falls	Laupāhoehoe	South Point	Kehena	Kailua-Kona	Kohala Resorts	Hilo Town
Chukar													
California Quail					U								
Gambel's Quail													
Japanese Quail													
Wild Turkey					U	C							
Common Peafowl					U	C							
Ring-necked Pheasant						U		U					
Chestnut-bellied Sandgrouse						U							
Eurasian Skylark						C		U					
'Elepaio	C												
'Ōma'o													
Palila													
'Ākepa													
Hawai'i 'Amakihi	C				C	C							
'Akiapōlā'au													
Hawai'i Creeper													
'Apapane	C				C								
'I'iwi					U								
Hwamei													
Red-billed Leiothrix	C				U								
Kalij Pheasant	U				U	U							
Cattle Egret			C	U									
Japanese White-eye	C	C	C	C	C	C	C	U		C	C	C	C
Spotted Dove			C	C	C	C	C	U	C	U	C	C	C
Zebra Dove			C	C	C	U	U	U	C	U	C	C	C

(continued)

Hawaiʻi Birds (2)

	Manukā	Whittington	Loko Waka Pond	ʻAimakapā Pond	Kaloko Drive	Puʻu Anahulu	ʻAkaka Falls	Laupāhoehoe	South Point	Kehena	Kailua-Kona	Kohala Resorts	Hilo Town
Mourning Dove						U							
Rock Dove					C	U					U	U	U
Common Myna	U	C	C	C	U	C	U	U	C	C	C	C	C
Northern Mockingbird						U		U			C		
Northern Cardinal	C	C	C	C	C	U	C	C		C	C	C	C
Yellow-billed Cardinal				C	U	U					C	C	
House Sparrow		U	C	U	C	U	U	C	U		C	C	C
House Finch		C	C	C	C	C	C	C	U		C	C	C
Saffron Finch				C	U	C					C	C	
Yellow-fronted Canary				U	C	C					C	C	
Red Avadavat				U	U						U		
Warbling Silverbill	U			U	U	C					C		
Java Sparrow				U	C	U					C	C	
Nutmeg Mannikin		U	C	C	C	U	U	C	U	C	C	C	C
Lavender Waxbill				U		U					U	U	

Maui Birds

	Hosmer/Waikamoi	Haleakalā	Polipoli	Keālia Pond	Kanahā Pond	Mākena Area	Wai'ānapanapa	Waihe'e Ridge	'Īao Valley	Kīpahulu District	Kapalua/Kā'anapali
Dark-rumped Petrel		U									
White-tailed Tropicbird		U							U	U	
Brown Booby							U				
Great Frigatebird				U		U	U				
Black Noddy							C				
Brown Noddy							U				
Black-crowned Night-Heron				C	C						C
Mallard					C						
Northern Pintail				C	C						
Northern Shoveler				C	C						
Lesser Scaup				U	U						
American Wigeon				U	U						
Green-winged Teal				U	U						
Hawaiian Coot				C	C						
Black-necked Stilt				C	C						
Bristle-thighed Curlew				U							
Long-billed Dowitcher				U							
Kōlea	C		C	C				C			C
Ruddy Turnstone				C	C						
Sanderling				C	C						
Wandering Tattler				C	C						
Nēnē		C									
Pueo	C	U	U					U			
Gray Francolin				U	U	C					C
Chukar	U	U	U								

(continued)

Maui Birds

	Hosmer/Waikamoi	Haleakalā	Polipoli	Keālia Pond	Kanahā Pond	Mākena Area	Wai'ānapanapa	Waihe'e Ridge	'Īao Valley	Kīpahulu District	Kapalua/Kā'anapali
Ring-necked Pheasant	C		C					U			U
Eurasian Skylark	C		C								
Hawai'i 'Amakihi	C		C					U			
Maui 'Alauahio	C		C								
'Apapane	C		C					U			
'I'iwi	C		C								
Hwamei	C		C					U	U	U	
Red-billed Leiothrix	U		C								
Cattle Egret				U	C						
Japanese White-eye	C		C	U	C	C	C	C	C	C	C
Spotted Dove	U			U	C	C	C	C	C	C	C
Zebra Dove				U	C	C	C	C	C	U	C
Rock Dove				U							
Common Myna	U	U		U	C	C	C	C	C	C	C
Northern Mockingbird		U	U	U		C					
Northern Cardinal	C		C	U	C	C	C		C	C	C
Red-crested Cardinal						C					C
House Sparrow	C					C	C			C	C
House Finch	C		C	U	U	C	C			U	C
Warbling Silverbill					U						
Nutmeg Mannikin				U	C	C	C	C	U		C

Moloka'i and Lāna'i Birds

	MOLOKA'I	Kamakou	Kakahai'a	Kaluako'i Resort	Kapuāiwa Grove	Palā'au Park	Moa'ula Falls	Kaluaapuhi Pond	LĀNA'I	Keōmoku Road	Lāna'ihale	Mānele Bay Area	Kaunolū
White-tailed Tropicbird	C					C					C		
Brown Booby												C	
Great Frigatebird		U	C	U	U		U			U	U		
Black-crowned Night-Heron		U					U	U					
Hawaiian Coot		U	U					U					
Black-necked Stilt		U		C									
Kōlea		C	C	C						C	C		
Ruddy Turnstone			C	U									
Sanderling			U	U									
Wandering Tattler			U	U									
Pueo		U	U								U	U	
Barn Owl		U											
Black Francolin		C	C		C	C							
Gray Francolin		C	C		C	C				C	C	C	C
Erckel's Francolin										C	C	C	C
California Quail			C										U
Gambel's Quail													U
Wild Turkey			U							U	U	U	
Ring-necked Pheasant		U	U					U		U	C	C	U
Eurasian Skylark		U	U					U					
Hawai'i 'Amakihi	C												
'Apapane	C										U		
'I'iwi	U												
Japanese Bush-Warbler	C										C		
Red-billed Leiothrix	C												

(continued)

Molokaʻi and Lānaʻi Birds

	MOLOKAʻI	Kamakou	Kakahaiʻa	Kaluakoʻi Resort	Kapuāiwa Grove	Palāʻau Park	Moaʻula Falls	Kaluaapuhi Pond	LĀNAʻI	Keōmoku Road	Lānaʻihale	Mānele Bay Area	Kaunolū
Cattle Egret		C											
Japanese White-eye	C	C	C	C	C	C	C			U	U	U	U
Spotted Dove		C	C	C	U	C	C			U	U	C	U
Zebra Dove		C	C	C	U	U	C			C	U	C	U
Common Myna		C	C	C	U	C	C			C	U	C	C
Northern Mockingbird		U	C	U				U		C	U	C	U
Northern Cardinal	U	C	C	C	U	U				C	C	C	U
Red-crested Cardinal		U	C	C						C		C	
House Sparrow		C	C	C	U							C	
House Finch		C	C	C	U			C		U		U	U
Warbling Silverbill		U								U	U		
Nutmeg Mannikin		C	C	C				U		C	C	C	U

Organized Hiking

Many groups in Hawai'i lead organized hikes. These groups can gain access to areas that are closed to individuals, so hiking with them is a good way to visit birding areas that would otherwise be off-limits. The majority of hikes are on O'ahu, but some are scheduled on other islands. Audubon Society trips generally emphasize birds. Other groups may focus on plants, archaeology, or some other aspect of an area. Some groups charge a small fee, usually a dollar or two, for trips.

The Hawai'i Audubon Society (HAS) usually offers one field trip per month, generally on the third Sunday. Most trips are on O'ahu. An annual calendar of trips is usually published each January. Trips are sometimes rescheduled due to weather or other factors. HAS supports many activities related to the protection of Hawai'i's native wildlife, such as wetland protection, alien species control, watershed preservation, and academic scholarships. A monthly journal, the *'Elepaio*, publishes information on birds of Hawai'i and the South Pacific.

The Hawai'i Audubon Society
850 Richards Street, Suite 505
Honolulu, Hawai'i 96813
(808) 528-1432

The Hawaiian Trail and Mountain Club organizes many hikes, usually on O'ahu and usually on weekends, often more than one hike per weekend. Send $1.00 and a self-addressed stamped envelope, and request a quarterly Schedule of Events.

The Hawaiian Trail and Mountain Club
P.O. Box 2238
Honolulu, Hawai'i 96804

The Nature Conservancy of Hawai'i protects over 17,000 acres on five islands through management of lands that are purchased or secured through conservation easements. Several of their preserves are open to the public, and two are of special interest to birders: Kamakou on Moloka'i and Waikamoi on Maui. The best way to see them is on guided hikes that are offered every month. For general information on programs and preserves, contact the office in Honolulu. To reserve space on hikes, contact the preserve managers.

The Nature Conservancy of Hawai'i
1116 Smith Street, Suite 201
Honolulu, Hawai'i 96817
(808) 537-4508

The Nature Conservancy of Hawai'i
Waikamoi Preserve Manager
P.O. Box 1716
Makawao, Maui, Hawai'i 96768
(808) 572-7849

The Nature Conservancy of Hawai'i
Kamakou Preserve Manager
P.O. Box 220
Kualapu'u, Moloka'i, Hawai'i 96757
(808) 553-5236

The Sierra Club leads hikes or camping trips nearly every weekend. Outings are scheduled on O'ahu, Maui, Kaua'i, and the Big Island. For a free copy of the quarterly outing schedule, call or write the Sierra Club office in Honolulu. The Sierra Club works actively for the protection of the environment and responsible use of natural resources through programs including public education, lobbying, and legal actions.

The Sierra Club
P.O. Box 2577
Honolulu, Hawai'i 96803
(808) 538-6616

The Hawai'i Nature Center (HNC) is a nonprofit organization that promotes wise stewardship of the Hawaiian Islands, largely through interpretative programs for schoolchildren. The HNC also leads hikes, some geared for adults and some for children *(keiki)*. A calendar of hikes is included in a quarterly newsletter, *The Steward*. Send $2.00 to get a copy, or call the center for information. The HNC also offers free trail maps of the Makiki Valley area where the Nature Center is located, and has modest interpretive displays open to the public at the center (see Makiki Valley description in Chapter 4). The HNC has opened a second center, with interpretive displays and gift shop, in the 'Iao Valley on Maui.

The Hawai'i Nature Center
2131 Makiki Heights Drive
Honolulu, Hawai'i 96822
(808) 955-0100

Government Agencies

THESE AGENCIES are valuable sources of information and assistance. They can provide you with maps, trail guides, hiking and camping permits, cabin rentals, and friendly free advice.

The State Department of Land and Natural Resources (DLNR) offers many services of interest to birders. Within the DLNR, the Division of Forestry and Wildlife has the impossible task of trying to please everyone who is interested in Hawaiian forests and wildlife. Hawai'i's 11,000 hunters (about 1 percent of the state's population) want numbers of sheep, pigs, and goats kept high so hunting will be good, while people concerned with the preservation of native plants and protection of watersheds want feral animal populations kept very low. Those who appreciate Hawai'i's native forests—the only tropical forests in the United States—for their value to wildlife want these forests protected. Other people, who make their living through logging or cattle ranching, want the forests managed for uses that are not always compatible with the needs of wildlife. Forestry and Wildlife offers a variety of trail maps and literature on plants and animals of Hawai'i. On the Big Island, where birders will need permits to enter certain areas, both the Baseyard in Hilo and the Tree Nursery outside Waimea (Kamuela) can issue them.

Department of Land and Natural Resources
Division of Forestry and Wildlife
1151 Punchbowl Street
Honolulu, Hawai'i 96813
(808) 587-0166

Department of Land and Natural Resources
Division of Forestry and Wildlife Baseyard
19 East Kawili Street
Hilo, Hawai'i, Hawai'i 96720
(808) 974-4221

Department of Land and Natural Resources
Division of Forestry and Wildlife Tree Nursery
Kamuela, Hawai'i, Hawai'i
(808) 887-6061

Department of Land and Natural Resources
Division of Forestry and Wildlife
54 South High Street
Wailuku, Maui, Hawai'i 96793
(808) 984-8100

Department of Land and Natural Resources
Division of Forestry and Wildlife
3060 Eiwa Street
Līhu'e, Kaua'i, Hawai'i 96766
(808) 274-3433

The Division of State Parks issues camping permits and rents cabins in State parks. In a state where lodging can be very expensive, State parks offer bargain-priced accommodations in some of the most scenic natural areas.

Department of Land and Natural Resources
Division of State Parks
P.O. Box 621
1151 Punchbowl Street
Honolulu, Hawai'i 96809
(808) 587-0300

Department of Land and Natural Resources
Division of State Parks
P.O. Box 936
75 Aupuni Street
Hilo, Hawai'i, Hawai'i 96721
(808) 974-6200

Department of Land and Natural Resources
Division of State Parks
54 South High Street, Room 101
Wailuku, Maui, Hawai'i 96793
(808) 984-8109

Department of Land and Natural Resources
Division of State Parks
P.O. Box 1671
3060 Eiwa Street
Līhu'e, Kaua'i, Hawai'i 96766
(808) 241-3446

The Natural Area Reserves System manages reserves such as Ka'ena Point on
O'ahu. Contact this office for a copy of the Ka'ena brochure or other information.
Department of Land and Natural Resources
Natural Area Reserves System
Kendall Building
888 Mililani Street
Honolulu, Hawai'i 96813
(808) 587-0054

The U.S. Fish and Wildlife Service manages over a dozen national wildlife refuges
(NWR) in Hawai'i and the tropical Pacific. The Hawaiian Islands NWR, established in 1909 by Teddy Roosevelt, protects 14 million seabirds of eighteen species on the Northwestern Hawaiian Islands. It has limited public access. There
are six wetland refuges in Hawai'i: Hanalei and Hule'ia on Kaua'i, Kakahai'a on
Moloka'i, Keālia Pond on Maui, and James Campbell and Pearl Harbor on O'ahu.
The most accessible are Hanalei (see Chapter 5) and Keālia (see Chapter 7).
Hule'ia will need extensive modification before it is attractive to endemic water-
birds. Kakahai'a can be viewed from a distance (see Chapter 8). James Campbell
is open by special use permit and occasional tours, and there is viewing at adjacent aquaculture ponds in Kahuku (see Chapter 4). Kīlauea Point NWR protects
seabirds on Kaua'i (Chapter 5). Finally, Hakalau Forest NWR on the Big Island is
the only NWR in the nation that has been established for the protection of forest
birds (Chapter 6).
U.S. Fish and Wildlife Service
Pacific Islands Office
300 Ala Moana Boulevard, Room 5302
Honolulu, Hawai'i 96813
(808) 541-1201

U.S. Fish and Wildlife Service
James Campbell National Wildlife Refuge
66590 Kamehameha Highway, Room 2C
Hale'iwa, Hawai'i 96712
(808) 637-6330

U.S. Fish and Wildlife Service
Kaua'i National Wildlife Refuge Complex
P.O. Box 87
Kīlauea, Hawai'i 96754
(808) 828-1413

U.S. Fish and Wildlife Service
Hakalau Forest National Wildlife Refuge
154 Waianuenue Avenue, Room 219
Hilo, Hawai'i 96720
(808) 933-6915

The National Park Service manages several facilities in Hawai'i, including two where hikes or interpretive nature programs of interest to birders are offered: Hawai'i Volcanoes National Park on the Big Island and Haleakalā National Park on Maui (see Chapters 6 and 7).

Hawai'i Volcanoes National Park
Hawai'i National Park, Hawai'i 96718
(808) 967-7311

Haleakalā National Park
P.O. Box 369
Makawao, Maui, Hawai'i 96768
(808) 572-9306

The Hawai'i Commission on Persons with Disabilities sells a five-part guidebook with information regarding accessibility of hotels, points of interest, beaches, and so forth.

Commission on Persons with Disabilities
919 Ala Moana Boulevard, Suite 101
Honolulu, Hawai'i 96813
(808) 586-8121 (Voice and TDD)

The Maui County Parks and Recreation Office in Kauanakakai can issue camping permits for Pāpōhaku Beach Park. Permits can be obtained in advance by mail or in person at the office in Kaunakakai. Permits cost $3.00 per night per person.

County Parks and Recreation Office
90 Ainoa Street
P.O. Box 1055
Kaunakakai, Hawai'i 96748
(808) 553-3204

Glossary

'a'ā: A type of lava characterized by rough, jagged surfaces. See *pāhoehoe.*

ahu: A cairn or rock marker, often used to mark trails. These are particularly useful where trails cross lava flows.

alien: A species that is not native to an area, but has been brought by people; introduced or exotic.

canopy: The tallest layer of vegetation in an area, usually the tallest trees.

Diamond Head: 1. Popular name for the extinct volcanic crater at the east end of Waikīkī in Honolulu. 2. A direction of travel in Honolulu and vicinity, toward the volcanic crater of the same name.

endemic: Evolved in a particular area and restricted to that area; found nowhere else. The 'I'iwi and many other honeycreepers are endemic to Hawai'i.

Estrildid: A bird of the family Estrildidae; small, often colorful finches commonly kept as cagebirds. Many have been introduced to Hawai'i.

'Ewa: A direction of travel in Honolulu and vicinity that is away from Diamond Head.

exclosure: A fence or other barricade meant to keep animals out of an area. Rare plants in Hawai'i are often protected from grazing animals with exclosures.

exotic: A species that is not native to an area, but has been brought by people; introduced or alien.

feral: An animal of a domestic species that has returned to the wild.

indigenous: A species occurring naturally in an area but not restricted to that area. The Black-crowned Night-Heron is indigenous to Hawai'i.

keiki: Hawaiian word for child.

kīpuka: Hawaiian word for a patch of vegetation that is surrounded by newer lava flows that have less, or no, plant life. Kīpuka are often called islands of vegetation.

koa: An important tree species in Hawaiian forests, especially moist forests at middle and upper elevations. Its scientific name is *Acacia koa.* Koa trees support many insects that provide food for birds. Only very young trees have true leaves; mature plants bear flattened, sickle-shaped leaf stems called phyllodes.

leeward: Generally to the southwest, opposite in direction from prevailing trade winds. See windward.

lehua: The blossom of the 'ōhi'a tree. An important source of nectar for Hawaiian birds.

lichen: A type of plant consisting of a fungus (plural: fungi) and an alga (plural: algae) that have a mutually beneficial relationship called symbiosis. The alga photosynthesizes, while the fungus provides the structure of the plant.

makai: Hawaiian word for toward the sea. See *mauka.*

māmane: A small tree of high, dry forests. Its scientific name is *Sophora chrysophylla.* The green seedpods of the māmane are an important food source for the Palila.

mauka: Hawaiian word for toward the mountains. See *makai.*

mesic: Fairly moist. Used to describe an area in terms of its rainfall (50 to 75 inches per year). Between dry and wet areas.

naio: A small tree of dry forests that often grows with māmane. Its scientific name is *Myoporum sandwicense.* The berries of the naio provide food for Palila and introduced birds.

native: Occurring naturally in a place, rather than introduced by people. Native species may be endemic to a location, or they may be indigenous.

'ōhi'a: The most prominent tree in Hawai'i, sometimes called 'ōhi'a-lehua. Its scientific name is *Metrosideros polymorpha.* The 'ōhi'a grows from sea level to over 8,000 feet, and is often the first plant to grow in new lava flows. The blossoms of the 'ōhi'a attract many Hawaiian birds.

pāhoehoe: A type of lava characterized by smooth surfaces, with a texture sometimes described as ropy. See *'a'ā.*

Passerine: A bird that is a member of the order Passeriformes, sometimes called perching birds or songbirds.

pavilion: A mostly open shelter for protection from the elements, especially rain. Often found in Hawaiian parks.

understory: The layer of forest vegetation that is below the canopy.

windward: Generally to the northeast. The northeast coast of a Hawaiian island is called the windward side because it faces the prevailing trade winds.

Additional Reading

Hawaiian Birdlife, Andrew J. Berger, University of Hawai'i Press, 1972, second edition 1981. A detailed review of island birdlife that is a readable mix of scientific information and anecdotes drawn from Berger's vast experience. Even the second edition has become somewhat dated.

Reference Maps of the Islands of Hawai'i, James A. Bier, University of Hawai'i Press. Separate maps cover the islands of O'ahu, Kaua'i, Hawai'i, Maui, and Moloka'i-Lāna'i. Each one is in full color and shows topography, roads, parks, some trails, and points of interest. These are the best maps for getting around in the Islands. Available from Marketing Department, University of Hawai'i Press, 2840 Kolowalu Street, Honolulu, HI 96822, (808) 956-8255.

Hawaii Handbook, J. D. Bisignani, Moon Publications, fourth edition 1995. In a sea of travel guides that are all pretty much the same, here's one that's different. It covers all the attractions, but places more emphasis on nature, small businesses, great hole-in-the-wall restaurants.

Hawaiian Hiking Trails, Craig Chisholm, 1994. Order from The Fernglen Press, 473 Sixth Street, Lake Oswego, Oregon 97034, (503) 635-4719. $15.95 postpaid. This handy volume describes fifty hikes on six islands, complete with trail maps.

Islands in a Far Sea: Nature and Man in Hawaii, John L. Culliney, Sierra Club Books, 1988. This exhaustive review of the biology of Hawai'i, and the usually negative effects that humanity has had, is packed with fascinating information.

Hawaii: The Islands of Life, Gavan Daws, The Nature Conservancy of Hawai'i, 1988. The ultimate coffee-table book for the lover of Hawaiian nature.

Shoal of Time: A History of the Hawaiian Islands, Gavan Daws, University of Hawai'i Press, 1968. A very readable history of the islands.

Seabirds of Hawaii, Craig S. Harrison, Cornell University Press, 1990. A thorough yet readable discussion of Hawai'i's marine birds, including an extensive section on conservation.

Hawaii's Birds, Hawai'i Audubon Society, fourth edition 1993. A small field guide to Hawaiian birds that is filled with color photographs. A must for birders in Hawai'i. $11.50 postpaid from the Hawai'i Audubon Society.

Maui Trails—Walks, Strolls, and Treks on the Valley Isle, Kathy Morey, Wilderness Press, 1991. This excellent book describes hiking opportunities at many Maui birding spots. The descriptions of trails at Polipoli State Park are especially helpful.

Birds of Hawaii, George C. Munro, Bridgeway Press, 1944. Reprinted by Charles E. Tuttle, Co. A fascinating account of Munro's observations, stretching from 1890 to World War II, on the status of each Hawaiian species.

A Field Guide to the Birds of Hawaii and the Tropical Pacific, H. Douglas Pratt, Phillip Bruner, and Delwyn G. Berrett, Princeton University Press, 1987. An excellent and detailed field guide to birds of Hawai'i and islands to the south and west.

Enjoying Birds in Hawaii: A Birdfinding Guide to the Fiftieth State, H. Douglas Pratt, Mutual Publishing Co., 1993. This detailed book provides the information a dedicated life-lister will need to spot as many species as possible.

Voices of Hawaii's Birds, H. Douglas Pratt, Hawai'i Audubon Society, 1996. This two-cassette set of recordings will help you identify the birds you hear but can't see. Very helpful for avoiding frustration in the forest. $27.35 postpaid from the Hawai'i Audubon Society.

New Pocket Hawaiian Dictionary, Mary Kawena Pukui, Samuel H. Elbert, Esther T. Mookini, and Yu Mapuana Nishizawa, University of Hawai'i Press, 1992. Great for looking up place-names, phrases, and so forth.

Checklist of the Birds of Hawaii—1992, Robert L. Pyle, Hawai'i Audubon Society. Lists all species naturally occurring in Hawai'i, and introduced species that have established viable populations. Gives status for each species and recent changes in status. $3.00 postpaid from the Hawai'i Audubon Society.

Fieldcard of the Birds of Hawaii, Robert L. Pyle and Andrew Engilis, Jr., Hawai'i Audubon Society, 1987. A pocket-sized field card listing bird species found in Hawai'i with space for notes of field trips. $1.25 postpaid from the Hawai'i Audubon Society.

All About Hawaiian, Albert J. Schütz, University of Hawai'i Press, 1995. A factual sketch of the language with a pronunciation guide and helpful word lists.

Forest Bird Communities of the Hawaiian Islands: Their Dynamics, Ecology, and Conservation, J. Michael Scott, Stephen Mountainspring, Fred Ramsey, and Cameron B. Kepler, 1986. Order from Western Foundation of Vertebrate Zoology, 439 Calle San Pablo, Camarillo, CA 93012–8506, $26.50 postpaid. In the mid-1970s biologists recognized the need for good baseline data on the birds of Hawai'i, so that subsequent conservation success or failure could be measured. For the next eight years a dedicated team of researchers surveyed and analyzed bird populations, resulting in this incredibly detailed look at forest birds. A must for the really serious birder. Sadly, some species have declined markedly since the survey was completed.

Oceanwatcher, Susan Scott, Green Turtle Press, 1988. This above-water guide to marine animals of Hawai'i is loaded with color photos and information.

Conservation Biology in Hawaii, edited by Charles P. Stone and Danielle B. Stone, distributed by University of Hawai'i Press, 1989. A very understandable primer on aspects of conservation biology from soils to terrestrial invertebrates.

Mark Twain's Letters From Hawaii, Mark Twain, University of Hawai'i Press, 1975. Tourism in Hawai'i was a bit different in 1866. You can't always take Twain at face value, but this volume gives a humorous glimpse of an earlier time.

Hawaii, Naturally, David Zurick, Wilderness Press, 1990. An island-by-island guide to environmental and cultural attractions.

Index

Note: Bird species are listed in the index only where the text provides some detail. To find all the sites where a bird occurs, use the occurrence tables. Color plates follow page 84. Italicized page numbers refer to black-and-white photographs.

Addendum

The following notations update information in the first printing as of 1998.

Chapter 2: A Hawaiian Natural History

p. 20   Recent analysis of honeycreeper DNA suggests these birds' ancestors may have arrived in Hawai'i as little as 3.5 million years ago, not the 15 to 20 million years previously hypothesized.

Chapter 4: O'ahu

p. 37   Kapi'olani Park. The Nutmeg Mannikin, formerly abundant on all main islands, has been declining rapidly in some areas over the past few years. It is now uncommon at Kapi'olani Park and elsewhere on O'ahu. Two other birds with similar habits, the Common Waxbill and Java Sparrow, have become more common.

p. 41   Makiki Valley. The Red-billed Leiothrix was common in O'ahu forests 20 or 30 years ago but then declined markedly. Now this bird is increasing again; look for it above Makiki Valley and along other forest trails. Common Waxbills are also increasing; they can be seen on sunny lawn areas in Makiki Valley where they feed on grass seeds.

p. 45   'Aiea Loop Trail. Freeway signs on the way to this site have changed. To reach the area from Honolulu, take the H-1 Freeway west ('Ewa) until it splits, forming H-1 and Route 78. Stay on Route 78, past the "Stadium/Halawa/Camp Smith" exit. Keep far to the left on Route 78 toward "Aiea/Pearlridge" and take the "Stadium/Aiea" exit. Then proceed as described on p. 45.

p. 51   Kawainui Marsh. The roadside area near Ka'elepulu Stream no longer offers good birding. Visit the Campbell National Wildlife Refuge instead (see below).

p. 58   Kahuku Area Wetlands. The Campbell National Wildlife Refuge now offers regularly scheduled guided tours Thursday afternoons from 4:00 to 6:00 P.M. and two Saturdays each month. Tours are scheduled from August or September through February 15, when resident Hawaiian Stilts are not nesting. To reserve tour space call the refuge office up to two months in advance at (808) 637-6330.

Chapter 5: Kaua'i

p. 77   Kōke'e State Park. Nēnē introductions in this area have been successful. Look for these geese around the museum and at the lookouts.

p. 92   Wailua River. The Nutmeg Mannikin has declined sharply here and elsewhere on Kaua'i. This bird is now uncommon. For a time, numbers of the Chestnut Mannikin increased and this bird seemed to be displacing the Nutmeg Mannikin on Kaua'i, but now the Chestnut, too, is uncommon. These population shifts have occurred very rapidly, in just a few years' time.

p. 93   Mānā Ponds. An old sand mine near Mānā is being restored as wetland habitat. Look for the Kawai'ele Sand Mine Bird Sanctuary at the 30.7-mile point along Highway 50.

p. 95   Kōloa Area. Sugar cane is being replaced by crops such as corn around Kōloa, and this is attracting Rose-ringed Parakeets that had become established near Kalāheo and Hanapēpē.

p. 97   Kaua'i Marriott Resort. This resort no longer offers wildlife boat tours but Nēnē can still be seen on the hotel grounds and golf course.

Chapter 6: The Big Island

p. 112   Saddle Road. The cabins at Mauna Kea State Recreation Area, mile marker 35.0 on the Saddle Road, have been closing periodically due to dwindling water supplies. Even if you can't rent a cabin, the area is worth visiting. Look for large coveys of California Quail around the cabins, especially in the morning.

Chapter 7: Maui

p. 130   Loko Waka Pond. Parking is now prohibited on the shoulder of the road adjacent to the pond. Use the county park across the road instead.

p. 147   Polipoli Springs. The road to Polipoli from Highway 377 is now marked as Waipoli Road.

Chapter 8: Moloka'i

p. 169   'Ō'ō'ia and Kaluaapuhi Ponds. Vegetation now blocks views of these ponds. A better choice for viewing wetland birds is the Kaunakakai Sewage Treatment Plant settling ponds. Go just 0.2 miles west of the zero-mile marker on Highway 460. Past the bridge over Kaikai Stream, but before the paved entrance to the treatment plant, is a wide dirt path that leads to the edge of the ponds. Also on Moloka'i, a nonprofit group has established a facility for captive breeding of Nēnē and for related environmental education projects. Visiting birders are welcome to call Nēnē O Moloka'i at (808) 553-5992 to arrange a visit.

About the Author

RICK SOEHREN lives in northern California and makes frequent birding trips to Hawai'i. He holds a B.S. degree from the University of California at Davis. A biologist active in conservation and resource management issues in Hawai'i and California, he is employed as an environmental specialist by the California Department of Water Resources. He is the author of numerous articles and government documents.